Man of the Trees

Richard St. Barbe Baker
O.B.E., Hon. LL.D. (Sask.),
F.I.A.L., For. Dip. Cantab.

1889 – 1982

Man of the Trees

Selected Writings of Richard St. Barbe Baker

Edited by
Karen Gridley

Ecology Action
Willits, California

Ecology Action
5798 Ridgewood Road
Willits, California 95490 U.S.A.
Printed in the United States of America by
Braun-Brumfield, Ann Arbor, Michigan

© Copyright 1989 by Ecology Action of the Midpeninsula.
All rights reserved.

Quotes reprinted with the kind permission of the Literary Trustees of the estate of the late Richard St. Barbe Baker and the Richard St. Barbe Baker Foundation.

Ecology Action is a small non-profit organization dedicated to learning, living and sharing human-sized sustainable ways of life on the earth.

Cover illustration by Lorraine and Stan Phillips.

Frontispiece portrait by Kay Holmes Stafford, courtesy of Mother Earth News.

ISBN 0-9600772-0-0

ISBN 0-9600772-1-9 (Paperback)

Ye are the fruits of one tree,
and the leaves of one branch.

– Bahá'u'lláh

Contents

Foreword by Charles J. Lankester   1

Preface   5

Introduction   7

1. Woodland Rebirth   17

2. Forest Community   21

3. Sylvan Heritage   35

4. Trees and Life   45

5. Skin of the Earth   57

6. Ancient Groves   69

7. Desert Challenge   79

8. History Repeats   91

9. New Earth   105

Sources   117

Foreword

Charles J. Lankester

Just a few days before I was asked to foreword this book about Richard St. Barbe Baker, several millions had awoken at various times on the sixteenth of July 1989 to read the Declaration of the 1989 Economic Summit which had just concluded in Paris. For the first time global environmental issues had been prominent on the agenda, indeed 18 of the 56 paragraphs of the final communiqué were devoted to these matters. We read that "deforestation also damages the atmosphere and must be reversed" and that "preserving the tropical forests is an urgent need for the world as a whole." These issues, we learned, had been the subject of keen debate by world leaders at the Economic Summit, and though the communiqué was longer on rhetoric than substance, I am nonetheless certain that somewhere St. Barbe is gently smiling. The environmental damage he had crusaded to bring to the attention of the many statesmen of his time was finally on the table, indeed the leaders in Paris stated "that in order to achieve sustainable development we shall ensure the compatibility of economic growth and development with the protection of the environment." Never one to remonstrate against those who failed to share his own farsighted vision of development, one that was harmonious with the natural environment, we can nonetheless safely bet that St. Barbe has murmured "better late than never" during the past couple of weeks.

What more appropriate and timely tribute to the centennial of the birth of this remarkable man could there be than the Paris communiqué. There is now broad recognition that, in future, environmental factors must be an integral part of economic decision making, and that good environmental and economic policies are mutually reinforcing. At the same time it is tragic that St. Barbe's warnings were not heeded much earlier. Now we must all bear witness to the premonitions of this pioneering environmentalist to the very real threats we have created to the future of our planet.

I had the good fortune to know St. Barbe. I met him for the first time in 1962 at the Headquarters of the Food and Agriculture Organization of the United Nations in Rome. He made regular pilgrimages there to plead with the staff of the Forestry Division for more development assistance to Forestry Departments throughout the developing world. This patient, purposeful and outspoken man was already well into his seventies and I suspect in hindsight that my superiors politely passed him on to me, the most junior recruit from British Columbia, because they didn't quite have the time or patience to handle him themselves. Obvious to the relief I was giving my colleagues, I enjoyed his various exploits beyond horizons I had only heard of and listened to his message that future generations of foresters must take up the environmental causes he had pioneered. We had many good chats in those surroundings over the next eight years as well as at World Forestry Congresses, when he invariably took time to inquire about the progress of the various projects I had been assigned. He was in his late eighties when he addressed his final World Congress. He scolded the participating economists to worry less about rates of return and more about the health of the forests they were managing. His final words on that occasion as the old man of the trees were to encourage all the delegates to join his reforestation crusades.

Many thought St. Barbe eccentric, but take a moment to reflect on today's most pressing environmental crises: the loss of some 17 million hectares to desertification and zero productivity each year, food security and the grim forecast that 65 countries may not be able to feed themselves by the year 2000; the need for agroforestry practices to conserve the precious topsoil in agricultural areas and to maintain the hydrological cycle by improving watershed management; the lost economic opportunities for future generations resulting from the accelerating loss of biodiversity;

and especially the menace of climate change. You will find all these issues mentioned in the following splendidly chosen quotations from his many books. Even more remarkable to me were his efforts to divert massive expenditures on armaments to conservation activities (page 88) and his solemn warning about the importance of conserving the Amazonian rainforests and the need "to develop a scheme whereby those presently in control of the Amazon can be compensated by the rest of the world for the forests they leave intact." In this and other passages we find St. Barbe both recognizing that the developed nations are responsible for most of the environmental destruction, and that while each developing nation has a sovereign right to manage its own resources, there is an urgency for all the nations to work in concert together to maintain the health of our global village.

This collection of quotations has also reminded me of St. Barbe's deep religious beliefs, his humility, sincerity, and his special concern for the less privileged and for our youth. These were the instincts that made him parry a European's blow to an African, an act which resulted in his dismissal from the colonial service, and that act alone speaks mountains for the decency and courage of St. Barbe.

This was a good and prophetic man. The centennial of his birth is fittingly celebrated with the strong resurgence of concern for his life-long beliefs that sustainable management of our forest resources is critical to our continued common well-being and perhaps to our very survival.

Charles J. Lankester
United Nations
Development Programme
New York, July 1989

Preface

In undertaking this book, Ecology Action set out to bring to a wider audience the ideas put forward by the late Richard St. Barbe Baker regarding our relationship to the natural environment, and in particular, to trees and forests.

The book is a selection of quotes drawn from seven of Dr. Baker's books: *The Redwoods, Green Glory, Sahara Challenge, Dance of the Trees, Land of Tané, Trees in the Environment,* and *My Life, My Trees*.

The quotes have been organized into nine chapters, each representing an essential theme related to Richard St. Barbe Baker's life and work. Each chapter starts with a paragraph outlining the theme to be addressed.

In editing the book, I have tried to let the quotes speak for themselves. Only an occasional italicized line has been inserted to introduce quotes that need to be placed in context. Where a particular word or phrase required explanation, an italicized editorial note follows in square brackets. The spelling used is, in most cases, as it appeared in the original text. The use of capital letters and punctuation, however, has been changed in many instances to conform with contemporary standards.

There are many people whose assistance made this book possible. I would like to particularly thank John Jeavons, Director of Ecology Action, and Hugh Locke, Literary Trustee of the Estate of Richard St. Barbe Baker, for their help in choosing and organizing the quotes. In addition, Robert White provided valuable assistance in writing the introduction. We are indebted to the Bahá'í International Community for underwriting the typesetting, design and layout of the book and cover. Bill Bruneau, Kathy Robison, and Robin Jeavons of Ecology Action and Leeta Gridley provided important assistance in preliminary editing and typing. J. Mogador Griffin was instrumental in maintaining the momentum of the original idea. We are grateful to Dean Nims and The Paper Corporation for their insightful support of this project.

Finally, we would like to thank the University of Saskatchewan Archives for having made available material from the Richard St. Barbe Baker Collection. A portion of the proceeds from the sale of this book will go to support research work on that collection. Anyone interested in the collection can write to: Richard St. Barbe Baker Collection, University of Saskatchewan Archives, Saskatoon, Saskatchewan S7N 0W0, Canada.

> Karen Gridley
> Willits, California
> July 1989

Introduction

Richard St. Barbe Baker, the Man

At four he planted two withies at the entrance to his garden — a withy being a willow twig — and they sprouted into trees. Perhaps it was this sort of impact on the physical world that inspired young Richard St. Barbe Baker to devote his life to the planting of trees. In his travels around the world, he became known as the Man of the Trees.

And yet it was not merely inspiration which guided the life of Richard St. Barbe Baker. Living from 1889 to 1982, St. Barbe — as he was known to his friends — dedicated his life to conservation forestry. This single-mindedness of purpose required and was matched by courage, perseverance, and a willing spirit of adventure.

His concern about tree cover for the planet led him to reforestation activities around the globe, starting first in Africa, as a colonial officer for the British government, and then on to sojourns in the Middle East, the United States, Canada, India, Australia and New Zealand, among many other countries.

Curiously, St. Barbe is little known to most Americans, despite his influence in protecting the California redwoods from as early as 1930, and his role in stimulating the governor of New York, Franklin Delano Roosevelt, into a plan of action for a future national Civilian Conservation Corps, now known as the *CCC*. Yet St. Barbe was known and admired by the British public and was an inspiration to foresters and conservationists of his time.

The Man of the Trees started out quietly in Hampshire, England, the son of a nurseryman and evangelical minister who fostered skills in growing gardens and trees, raising bees and relating to the physical world. From venturing into the pine forests near his country home, St. Barbe developed an almost ecstatic love for the wildness of nature and its beauty. During his school years, a particular old beech tree became his personal "Mother Confessor." The experience of harmony and trusting intimacy with nature shaped in St. Barbe a deeply personal sense of oneness that became the foundation for a life devoted to protecting nature.

Hearing the stories his father told him about his great uncle Richard, who had gone to Canada and shot a bear, St. Barbe longed as a young boy to follow his uncle's footsteps and experience adventure. And so he did, in 1910, with the intention of working his way through college in Saskatoon, Saskatchewan.

He developed a rather amazing ability to rustle up a livelihood with one skill or another, whether breaking broncos, working in lumber mills, or writing for the local paper. Often he was just scraping by from one meal to another. Or then he would turn some creative idea into a real moneymaker that would finance him through the completion of his goal.

It was in his years in Canada that St. Barbe began to be concerned about the trees and the land. While crossing the Canadian prairies, he first recognized a desert in the making. Vast stretches were ploughed up, and the grass and tree cover removed for the sowing of wheat and oats. Without protective cover the soil began to drift and blow away: up to an inch every year. St. Barbe would later recount the story of a Dorset man who had come to Saskatchewan to farm. He built his frame house, planted his fields, sowed his wheat and oats. After twenty years he decided the country was no good for farming, for eight feet of his soil had gone and he had to climb up into his house. He returned to Dorset, where he became a tenant farmer once again, with tree-surrounded fields. Later when St. Barbe was working in a lumber camp near Prince Albert, he became disheartened by the unnecessary waste of trees, and he determined to qualify for forestry. At this time, he began his practice of encouraging the planting of trees.

Although World War I broke out and borrowed a few of St. Barbe's years for the European battlefields, the deflection from his main love was only slight. Back in England, his goal of graduating from the School of Forestry at Cambridge was achieved despite his grumblings about the fact that one had to pass a searching examination in everything conceivable except "knitting and the care of teething infants!" After all, what he wanted was to get out into the forests and plant and care for trees. His college costs were paid by launching his own business. This time his scheme was to design mobile trailers that could be built out of surplus war aircraft material. Perhaps the degree of his ability to "act big" can be gauged by his purchasing trip to the Government Disposals Board. With the idea of experimenting on one caravan, he ventured a request for a couple of aeroplane undercarriages. The government agent exclaimed: "A couple! This is the Government Disposals Board. The smallest lot is thirty-six!" So St. Barbe took the lot!

It seems that St. Barbe's life as a professional forester and conservationist really began in 1920 when he went to Kenya under the colonial office as Assistant Conservator of Forests. Here he observed firsthand the stages of degradation from forest to desert: forests cleared and burned for farming, after which the soil was exposed to sun and rain and the humus quickly degraded. Without the forest to slowly absorb and store rainfall, water tables would eventually sink. St. Barbe also recognized that forests under colonial "conservatorship" could no longer be exploited without regeneration. The solution to the situation was obvious to St. Barbe: plant more trees.

The native Kikuyu, among whom St. Barbe worked, didn't seem interested in a tree planting program. "Trees," they said, "*shauri ya Mungu* — God's business." But St. Barbe sought to convince the Africans that if all of Mungu's parent trees were felled, there would be no young ones. He hit upon the ingenious idea of holding a Dance of the Trees. Recognizing the integral role of dance in the Kikuyu tribe, he offered a bullock as prize for the best-turned-out warrior and a beaded necklace for the most beautiful woman. Surprisingly, or perhaps not so surprisingly, three thousand warriors showed up for the event. From among these many warriors, he solicited volunteers who would be willing to become *Men of the Trees*, willing to commit to planting ten trees a year and protecting trees everywhere. This was social forestry, the guiding philosophy of current initiatives in development, long before the concept was invented.

Somehow, the combination of an encompassing vision and a respect, grounded in his appreciation and understanding of his

Kikuyu friends, enabled him to organize the Africans in a way quite unheard of. He was the first *pink-cheeked* to enter the Kikuyu secret society, and in turn, it was with the Kikuyu, that he founded the Men of the Trees, which became the focal organization for his lifelong efforts in forest conservation and earth regeneration.

In Africa, St. Barbe began to appreciate how the rhythm of growth was maintained in the virgin tropical forests — growth being balanced by decay — and how the forest was an interdependent, interrelated community of man, plants, animals, birds, insects, and fungi. Shying away from imported exotics, he began a tradition of planting and maintaining mixed native woods. Half a century later, the wisdom and far-sightedness of Richard St. Barbe Baker's mixed age, mixed species, native tree-farming practices have become obvious. Monoculture plantations throughout the world, planted all at the same time with a single species, have proven more vulnerable to insects and disease. In some instances, such as with non-native eucalyptus, tree stands have dried up streams and the water table, and have needed to be totally replaced.

St. Barbe's work in Kenya was followed in 1924 by a similar assignment in Nigeria, in the mahogany forests, where he explored the relationship of sound silvicultural practices and the economic demand for wood. His conservation measures did not preclude economic benefit, but he saw the necessity of preserving the *capital* of the forests in order to derive sustained benefits. St. Barbe considered clear-cutting as equivalent to skinning a person alive. He was a champion of a concept that is heard a lot today, namely, *sustained yield.* He also advocated the integration of forestry and agriculture as a way of maximizing land use while protecting the soil.

Because of his intervention on behalf of an African, (he parried a European's blow meant for one of his Nigerian men), he was discharged by the colonial office. Always one to turn seeming misfortune into opportunity, St. Barbe saw himself now free to pursue his personal vision of reforestation. This discharge coincided with a deepening involvement and then membership in the Bahá'í Faith, a factor which significantly influenced his world view and lent strength to his commitment to the work of the Men of the Trees and land reclamation.

Invited to Palestine in 1929, St. Barbe actually managed to get representatives of the major religions together to commit to a plan of reforestation of desert areas. The expansion of membership in the Men of the Trees was now assured. Tree nurseries popped up and tree plantings were begun.

From Palestine, all the signs pointed westward. From this period on, the life of St. Barbe reads like that of a somewhat eccentric and learned vagabond who traveled around the world planting trees and convincing everyone else to do the same — whether it was school children or Nehru or Franklin D. Roosevelt. A professional forester, a practical economist, a lecturer/writer, an inspiring conservationist, but most of all, someone willing to fly by the seat of his pants, St. Barbe would work his passage on a boat going to Australia or arrive in New York City with five dollars in his pocket, and make his first week's room and board by giving a talk, writing an article, or writing a book. His books are numerous and varied and most of them were written in a relatively short period of time. He dictated one in a period of ten days while lying flat on his back in bed.

Perhaps more than anything else, St. Barbe was a man of action. If a job needed doing, he'd do it. And he much preferred to show someone how to do something or do it with them, than to give an order. This quality endeared him to the many he had contact with in North America, New Zealand, Australia, Africa and India.

St. Barbe was not afraid of having his say and broadcasting it widely. Distraught by what was happening in New Zealand forestry while on a visit there, he called the newspapers together and gave his own little press conference, despite the fact that he was not yet well-known there.

Never deterred from his cause of reforesting and re-greening the planet, St. Barbe had an impact by dropping the right word at the right time. While at a ceremonial tree-planting in Ireland presided over by the President, he inquired of that dignitary what percentage of Eire was tree-covered, and then ventured the guess of two and one-half percent. Startled by the figure, the President queried his Minister of Lands right on the spot, to find that the actual figure was even less. St. Barbe's intervention led to an order to double the tree planting program.

Likewise Nehru — upon asking St. Barbe what to do about Indian deserts — was counselled to plant trees, to surround fields with trees, thus increasing production. Nehru followed St. Barbe's advice and increased production by as much as one hundred percent.

Based in England, Men of the Trees became a worldwide organization, sponsored in part by St. Barbe's lectures and talks. The group published a journal from as early as 1929, and a Tree Lover's Calendar. For several years, St. Barbe and his coworkers organized forestry summer schools held at different universities in England to stimulate an interest in conservation forestry.

When St. Barbe first saw the redwoods of California in 1930, he was awed beyond words. Rivaled only by the lordly kauri trees of New Zealand, the redwoods, he felt, should be preserved for all time.

Through his influence and in conjunction with Save-the-Redwoods League, he was able to obtain financial backing from several philanthropists. He launched a lecture tour in England about the *Wonder Trees of the World* and got support from tax-beleaguered Britains to save the redwoods.

His goal was to preserve an area of 12,000 acres, which he saw as essential to protect the delicate and complex life of the redwood forest. When a state park was established "for all time" in 1939, he felt he had accomplished his purpose. Only later, when he returned to Northern California in 1963, did he realize that there was no such thing as "preserving for all time." Each generation must fight its own fight.

Instigated by St. Barbe, The First World Forestry Charter Gathering was held in 1946. This forum was to provide an opportunity for the exchange of ideas between diplomatic representatives from many countries of the world. It was through this group that St. Barbe especially sought to tackle the reclamation of the Sahara Desert. Not discouraged by the size of the task, St. Barbe reasoned that if the United States could plant a thousand-mile windbreak across the western plains from Canada to Texas, and Russia could create a shelter belt the length of three thousand miles, then what would prevent the countries surrounding the Sahara Desert from collaborating on a shelter belt four thousand miles long and thirty miles wide?

St. Barbe felt strongly that the energy that was spent nationally and internationally on armaments and strife could find a creative outlet in a unified battle against the encroaching desert; the desert that was then advancing twenty-nine miles every year.

His efforts to mobilize a desert reclamation program, named the Green Front, lasted over many years. His first endeavor was a 26,000 mile exploratory trip across the desert in a secondhand desert Humber. This was perhaps the most exhilarating experience of his life. In his book *Sahara Challenge*, St. Barbe describes this incredible adventure, an adventure that began with his fanfare departure from Trafalgar Square. Here he was greeted by hundreds of Londoners bearing peach stones. These well-wishers had been alerted by a radio broadcast that the expedition team would plant them along their journey, as a first step in halting the desert.

The danger of traversing the deceptive sands of the Sahara created an excitement for the expedition that was only matched by the team's discovery of evidence that trees had once grown in these barren wastes. The high point for St. Barbe was at the well of Ekker, in the middle of the desert. There an old resthouse keeper described a great forest, called "the Last Forest," that had existed when he was a young boy. Convinced by his trip that the Sahara had once been wooded, at least in part, St. Barbe proceeded to influence colonial administrations of the feasibility of reclaiming the Sahara Desert.

As the years went by, definite reclamation achievements were made, and a Sahara Reclamation Program established, but the Sahara was now surrounded by twenty-five new African nations, struggling for political stability. Reforestation did not progress according to St. Barbe's vision. Yet he planted a seed that today is bearing fruit. Now reforestation programs are being mobilized internationally to halt the advancing Saharan sands. The fact is that St. Barbe was far ahead of the times in this, as in many other aspects of earth regeneration.

His later years, based at Mount Cook in New Zealand with his second wife Catriona, were spent in writing, traveling, organizing and passing on his message to school children of the world, in the hope that they would learn to love and protect their tree heritage and share his vision of earth healing.

St. Barbe, His Ideas

Decades before the earth was photographed from space and understood as a unitary ecological reality, St. Barbe was practicing forestry from this global and holistic perspective. His view of the planet as a living organism anticipated the Gaia theory, which has provided such a fruitful base for scientific investigation in recent years. Very early in his career, St. Barbe began to see the complex ecological interactions within the forest as a mirror of the organization of all life. He believed that life as a property of the whole ecosphere is maintained by the synergistic interrelationship of air, water, soil and organisms. From this vantage point, he saw forests as a vital organ within the self-regulating, life-sustaining whole. Knowing himself as part of this whole, he felt the urgency of protecting, conserving and planting trees. He felt we human beings must "play fair to the earth," to understand and serve this wholeness, which embraces all beings.

He also understood that ecological systems exhibit certain threshold responses. Thus removing too many trees over too large an area could dismantle a whole ecosystem and eventually disrupt the earth's capacity to maintain critical life-support functions. His views against clear-cutting reflected this understanding. Similarly, planting trees could institute a regenerative cycle and bring degraded ecosystems to a threshold of recovery. These ecological concepts were only beginning to emerge in science at the time St. Barbe was describing and practicing them. The world is now struggling with the consequences of ignoring them.

It is in this context that St. Barbe's writings can be best appreciated. He possessed a scientific, aesthetic, and spiritual perception of the forest all at once. To the reader, he gives an appreciation of the wonder, beauty, and sacredness of nature, while at the same time teaching ecological principles and our responsibility to understand and respect them. To combine these dimensions he often relied on metaphors and parables, as in the case of his description of the forest as the living planet's skin.

In the final analysis, St. Barbe lived his philosophy and it lived him. His life stands out as a model for a world desperately in need of the capacity to relate to the earth with respect, care, and responsibility. His life and his writings together demonstrate a vision of harmony between humanity and the earth. This is especially true of his efforts to tackle the Sahara Desert. The Sahara represented the largest man-made desert — an image of the destruction of the earth's bountiful life. The Green Front program against the desert symbolized the potential to bring people and nations into cooperation with each other, and with the forces of growth and healing in nature itself. It demonstrated a faith in the capacity of humanity to transcend the separatist, non-synergistic tendencies of its nature. In this sense, his life invites human beings to become a force in guiding the evolution of life. His own integration of science and religion led him to see the potential for developing a mature planetary civilization based on ecological and spiritual principles. Fulfilling this potential remains the challenge of our age.

1

Woodland Rebirth

"I had entered the temple of the woods."

Woodland Rebirth

No selection of quotes from Richard St. Barbe Baker would be complete without the description of his first woodland experience at the age of five or six. A deeply moving event, it set the scene for his lifetime championship of forests.

The surrounding woods were extensive and in those days I never penetrated very far, nor would Perrin take me into the forest as a rule. To a small boy their depths were mysterious and rather awe-inspiring. One day, I found I had exhausted my ideas of treescaping in the sandpit and so, greatly daring, asked if I might be allowed to go for a walk in the wood. Perrin said the woods were full of adders at this time of the year and no safe place for the likes of a small boy. But I coaxed her to let me go and reluctantly she allowed me to set out on what seemed to me a wonderful expedition. No explorer of space probing the secrets of other planets could have felt more exultation than I did at that moment.

As I set out on that greatest of all forest adventures, at first I kept to a path which wound its way down into the valley; but soon I found myself in a dense part of the forest where the trees were taller and the path became lost in bracken beneath the pines. Soon I was completely isolated in the luxuriant, tangled growth of ferns which were well above my head. In my infant mind I seemed to have entered the fairyland of my dreams. I wandered on as in a dream, all sense of time and space lost. As I continued this mysterious journey, looking up every now and then I could see shafts of light where the sunshine lit up the morning mists and made subtle shadows on the huge bracken fronds which provided a continuous canopy of bright green over me. Their pungent scent was a delight to me. Although I could see only a few yards ahead, I had no sense of being shut in. The sensation was exhilarating. I began to walk faster, buoyed up with an almost ethereal feeling of well-being, as if I had been detached from earth. I became intoxicated with the beauty around me, immersed in the joyousness and exultation of feeling part of it all.

Soon the bracken became shorter, and before long it was left behind as a clearing opened where the dry pine needles covered the floor of the forest with a soft brown carpet. Rays of light pierced the canopy of the forest, were reflected in the ground mists and appeared as glorious shafts interlaced with the tall stems of the trees; bright and dark threads woven into a design. I had entered the temple of the woods. I sank to the ground in a state of ecstasy; everything was intensely vivid — the call of a distant cuckoo seemed just by me. I was alone and yet encompassed by all the living creatures I loved so dearly.

As I lay back a dead twig snapped, like the crack of a whip; the birds warbling sounded like a cathedral organ. The overpowering beauty of it all entered my very being. At that moment my heart brimmed over with a sense of unspeakable thankfulness which has followed me through the years since that woodland rebirth.[1]

2

Forest Community

"A forest is a society of living things,
the greatest of which is the tree."

Forest Community

St Barbe's experiences reinforced his belief that a forest is not merely an aggregation of trees but rather a complex living community which requires an interdependence among man, plants, animals, birds, insects and fungi.

He saw in native vegetation a stability and resistance to disease that is disturbed by man's interventions.

In his own forestry practices throughout the world he sought to interfere minimally with nature and to emulate her principles whenever possible.

In a virgin forest, growth keeps pace with decay: at one place a tree will be reaching maturity, at another declining. As an old tree falls, a gap is made in the canopy, light is let in, and many younger trees race with each other to occupy this new opening. This is happening by slow degrees throughout the forest and so the rhythm of growth is maintained.

It has taken centuries to form the floor of the forest. It is possible that a forest of today may be standing on a site which at one time was bare rock. Perhaps some lichens established themselves on a damp level surface of the rock. The weathering action of the elements disintegrated the surface and formed a layer of tiny rock particles. This process was hastened by the chemical secretions of the lichens until these mosses were followed by grasses and in due order by herbs and shrubs, and then one day a wind-borne tree seed lodged there and began to grow, and thus founded a community of trees. That was how the forest began.

As the treed area developed, precipitation increased; growth was hastened as well as decay, acting and reacting upon the future generations of trees. In all this, fungi and bacteria played their part in producing and modifying humus and litter. Parent trees scattered their seed, which, borne by the wind, colonized fresh areas.

Each group of seedlings contended with the others for light and the suppressed and less fortunate in the race fed the growth of their victors.

There is continual decline as well as growth in the forest. There is no standing still: as soon as a tree stops growing it begins to decline. Even before this, decay may have set in within its heart, and it is set upon by beetles and borers, its roots are attacked by fungi, its huge body is no longer supported, and a mere gust of wind may bring it to earth. It becomes a home for woodlice and millions of other insects, which play their part in returning the fallen tree to the earth from which it sprang. [2]

The human body is the earth in miniature. For is it not true that each part of the body is dependent on the other? If any part becomes defective or is lost, the whole balance is upset. And so it is with the earth. For millions of years the earth has been built up, and maintained its unique routine which is now threatened. Let man go forward in partnership with nature and use his accumulated knowledge to foster and maintain this essential balance. [3]

When nature plants a forest, she mixes her species, the broad-leaved trees or hardwood and the coniferous trees producing softwood. They are different in form and foliage and the blending of their falling leaves and needles produces just the right texture of humus for plant growth. Some of these trees, such as beech, hornbeam and yew, are shade bearers and will be tolerant of the shadow cast by their neighbours.

Others, like silver birch, gean and larch, are light demanders and are intolerant of shade and so are easily suppressed by nearby trees. Some trees will stand many degrees of frost and others are less frost-hardy. Most trees need a well drained soil but there are some, like alders, that will survive in wet situations. Some, like willows and poplars, like to be near water but must have well drained soil. Who knows but that trees have as many likes and dislikes as human beings! [4]

Trees can give us so much, apart from things aesthetic and economic. They are forever giving out an element which is healthful and exhilarating. Fascinating as flowers may be, they do not give us the strength of a tree. A tree is a real thing, it has a personality. It is a living entity. Is it too much to believe that it responds to the love that is given to it?... A new and vital interest will arise in our planting activities when we come to realize that the objects around us are not inanimate, but very much alive. We shall tread softly when we enter the sanctuary of the woods, seeing we are in company with tree beings who respond to our love and care. A new world will open up to us, and in response to our love we shall gain a share of the vitality and inspiration that comes to a man or woman who is in love with trees. [5]

What happens underground is as important as on the floor of the forest or among the branches. If you could make an X-ray picture of a vertical section, you would find many different shaped roots — tap roots, support roots and myriads of tiny hair roots and feeders. Broadly speaking, the roots of trees fall into three main systems. There are the spear-shaped roots, the heart-shaped roots and the flat-surfaced roots.

The spear-shaped roots tap the minerals deep down in the ground, the heart-shaped help to lift the water, and the function of the flat roots is to support and feed a tree in wet situations. Trees are not only fed by their roots but breathe through them as they do through the lenticels in their bark. Sometimes, if the situation is very wet, the roots will form knees above the water so that the trees can breathe through these knees, like the swamp cypress.

The tideland or sitka spruce is a good example of a flat-rooted tree, the birch heart-shaped and the oak spear-shaped.

If you make a plantation all of one species, the roots compete with each other at the same level for food, moisture and support, and rob and weaken each other. [6]

On a visit to Scotland with Lord Sempill: For days we visited estates where trees had suffered as a result of the storm on January 31 that year, when a freezing wind had blown for eight hours, uprooting eleven million trees. We inspected the tree roots to read, if possible, the cause of this disaster.

We found that most of the trees felled in the storm were approximately forty years of age when root competition had become severe. It is a fact that the hair roots of pines are charged with an acid sheath; nature has provided this to help dissolve rocks and enable the root to penetrate. One often sees how the tiny root of a pine, by the sheer force of expansion, has succeeded in splitting a rock, emerging a foot or so below the point of entry.

Imagine myriads of small roots competing with each other at the same level for growing space. When this happens an acid pan is formed at the level of the greatest root competition. For the health of pines there must be a mixture of broad-leaved trees so that the leaf-fall can provide food for the roots of the conifers. Nature is wonderful in her adaptations for she provides a symbiotic fungus whose strands attach themselves at one end to the decaying leaf of the hardwood tree while the other end contacts the tiny feeding roots of the pine. This process is known as a micorrhizal association.

It was the outside trees of the plantations which had remained standing, while those inside were uprooted by the storm. This was because the trees on the outside enjoyed this micorrhizal association which arose from the leaves of the hedges and other non-coniferous plants and shrubs. [7]

The earth with its covering of soil, plant and tree is always in a process of evolution, for nothing is static in a society of living things. [8]

Speaking about certain tree species in the rainforests of Nigeria: Rarer and even more strongly aromatic are the *guareas*, which produce a darker, attractively grained wood. So much sought after is the guarea that a single log might fetch £2,000 in the Liverpool timber auctions. They are not gregarious, nor are they to be found in pure stands; an eloquent lesson to those wedded to monoculture in whatever form, in farm or forest.

Each mahogany is surrounded by numerous trees belonging to other families, amongst which is that important family of *Leguminoseae* — the soil improvers. These I have observed to be good nurse trees for the mahoganies. The more important species of mahogany require the services of a succession of nurse trees throughout their life to bring them to perfection. Some of these provide just sufficient competition to coax the young sapling upwards. Others do their work in secret under the surface of the soil, interlacing the roots, a sort of symbiosis, like the mycelium, which starts as an independent web-like growth, surrounds the sheath of plant rootlets and prepares food that can be assimilated by the growing trees. [9]

Our present knowledge of the native forest is so incomplete that it would be folly to fell indiscriminately without first assuring ourselves that it can be reconstituted as well as, or better than, it was originally. [10]

Referring to an area near Mt. Kenya: The lower foothills of the mountain are the domain of the mutarakwa, the pencil cedar, and mutamayu, the brown olive; not the edible kind, but a tree that has timber valuable for motor-bodies and now is being used in Kenya for the carved figures and penknives which are so popular. When on the following day I rode on a mule through this forest of juniper and olive, it was interesting to see how closely they are associated with each other. The pigeon feeds on the juniper berries while perching on the olive branches; moreover the juniper seed can only be germinated with the help of the gastric juices of this bird, who in discarding the seed in his droppings, lets it fall near the protecting olive. [11]

No farm is complete without its balanced home wood, where so many good influences are at work. No forest or home wood can contribute its best unless it contains a mixture of species, both broad-leaved and coniferous trees. Pure stands of conifers tend to make the ground acid and fail to provide litter with soil-improving properties. This may be accounted for by the fact that the micro-fauna of the soil is poorer under pines than under mixed hardwoods, and the number of earthworms is less and their weight but a third of those found in the soil of the natural woods. Moreover, pine woods are inhospitable to birds and wildlife. Lacking browse, deer are liable to bark the trees, and squirrels and grouse destroy the buds. When a pure coniferous forest is artificially created, the balance of nature is disturbed. Birds may forsake the area owing to absence of food and thus permit insect pests to establish themselves unmolested. Insectivorous birds are the friends of both the farmer and the forester and are well known to be important controls of farm and forest insects. [12]

It was King Canute, the Dane, who in the year 1018 was the first to set bounds and limits to those forests so that they might become *sanctuary* for game. [In Canute's words], "nothing so much defaces the forest as to cut down a covert which is vert *[green]* and for this and other causes the lawes provide to preserve it with great care. For 'tis the nature of wild beasts to frequent the coverts and great woods, where in the winter they are sheltered from the cold, and in the summer from the heat; and when they are hunted, they run thither *for protection*." [13]

On revisiting sites of his first conservatorship in Kenya: Next morning, we drove up together to my old forest station and the hill where thirty years before, three-thousand Kikuyu morans *[warriors]* had come together for the first Dance of the Trees. A lawn covered the site of my original nurseries, where the young men used to come in the evenings and prick out the little trees into boxes. There were still nurseries, though not in the same place; they were good little nurseries, but with too many pines and eucalyptus for my liking. There were also two patches of ground, both of which had been severely cut over and which were degenerating into scrub, where they were proudly experimenting with chemical fertilizers, as if they were the latest form of juju *[magic]*. Josiah took me through groves of eucalyptus trees which had been planted since my time and through them we looked across the valley where there had been a luxuriant virgin forest. It had nearly all gone. Only here and there were groves of eucalyptus. I was anxious to see how the native trees I had planted were doing; I wanted to find the mutarakwa and mutamayu which together we had planted. We had not far to go. We went down by a road right through the woods where we soon found our trees, now over a hundred feet high, with a natural forest floor of decomposing humus. No chemical fertilizers were needed here. It was a perfect forest, well cared for, having been periodically thinned. [14]

Referring to plantations in Kenya: [An] example of dangerous monoculture has been the foundation of pure plantations of Monterey cypress, or *Cupressus macrocarpa* and *Cupressus lusitanica*, in the high rainfall areas, especially where ground mists persist at certain times of the year. Many of these trees have become infected with canker caused by *Monochactia unicornis*, a fungus caused by water running down the stem and being caught in the niches... where the branches come out; this place provides ideal conditions in which the spore takes hold. The risk of infection is greatly reduced in a mixed plantation that includes a larger percentage of broad-leaf trees.

On seeing this excessive planting of exotics in monocultivated forests, it seemed to me that all my demonstrations and advice of thirty years ago had been forgotten.

I drove on through this unhappy country towards Mount Kenya. It was a relief to come to one of the rare remaining indigenous forests. Over it hung the most vivid rainbow I have ever seen. It looked as if my road went straight through the middle of the western arch, but it always kept ahead of me, seeming to move as I moved. At last it stood still and I drove right into a light shower of rain, and at that point the rainbow ended. I had come to the end too of the indigenous forest and had entered the eucalyptus world. It was the land where the rainbow ends. [15]

It is my belief that precipitation and water-holding qualities of the indigenous or natural forest are superior to that of any form of monocultivated exotic or introduced plantations. A contributory cause to the lowering of the water table is the continued clearing of the indigenous high forest and neglect to replace at all or to replant only with imported species.

It is well known that eucalyptus is good to plant in a swamp to dry it up, but a dangerous tree to grow extensively as cover in a land needing water. The balance of nature is so finely adjusted that it is only too easy to upset its equilibrium. The natural forest has adapted itself to soil and climate through the ages and is best suited to serve the biological requirements of the land. When indigenous forest of any country is opened up it should be carefully studied with a view to improving it, if possible by the introduction of silvicultural systems, before exploitation is begun. It is particularly important on the equator to repair the canopy of the high forest and never open it up too rapidly, otherwise growth is lost. [16]

It remains to be seen whether in the long run it is a good policy to sacrifice the native trees for fast-growing exotics. In monoculture

1 Richard St. Barbe Baker in the early 1930's.

2 St. Barbe with his first mobile trailer passing a gypsy caravan, 1919.

3   The inaugural ceremony marking the launch of the Men of the Trees in Kenya, 1922.

4   Five Kikuyu warriors, among the original members of the Men of the Trees.

5   A silvicultural experiment, involving selective tree felling, established by St. Barbe in Nigeria in 1927.

6   St. Barbe with a head man and three of his followers in Benin, Southern Nigeria.

7   Reunited with founding members of the Men of the Trees in 1953, with Chief Josiah Njonjo presiding.

8   Photograph taken to promote a lecture tour of the United States in 1929.

there is always a grave risk of attack by parasites or pests, which may sweep the whole area. Besides, these pure stands alter the character of the country, so that it no longer enjoys the serenity or dignity of the mighty forests which once cast their magic spell over the land. [17]

Careless burning of native bush on hillsides and steep gullies while clearing land for pastoral purposes has perhaps caused the most erosion, because no planting of exotics on hillsides can compare with the dense undergrowth of native bush to check the run-off of water and, at the same time, act as nurse-plants for seedlings.[18]

There are a large number of timbers of economic value, and the pleasing dramatic effects of many of the ornamental grained woods have yet to be fully appreciated.... The beauty of many of these woods, when skillfully converted, has to be seen to be believed.

Unfortunately, the virgin forests from which these trees come have been exploited with too little thought for future supplies. The lands which contain such valuable trees should be worked on a sustained-yield basis, so that supplies of their timbers may be available for all time. There is a danger that, when the virgin forests are felled, inferior species may take their place, for there has been a tendency to pick out the plums, or the trees that are most sought after for the beauty and usefulness of their wood, and leave the rest, with the consequence that the forest deteriorates, never to regain the natural composition which yielded those treasures. [19]

Upon visiting a new research station at the site of his original work in Kenya: I studied their reports on general forestry research, including forest pathology, where the canker disease of cypresses seems to have been their principal concern, a disease which should never have occurred under the proper silvicultural management in mixed woods.

It was not for me to assess the value of clove research or the gumming disease of coconuts, or to criticize their experimental work in insecticides for the Empire Cotton Growing Corporation, or the virus diseases of cassava, or the streak diseases of maize, or the suspected virus disease of sweet potatoes. So much of what I saw seemed to me to be obsolete, in view of the proven value of natural and ecological conditions. Today we are learning more and more that the best method of fighting disease is to provide the conditions in which it does not occur. If the health of a plantation, whether it be farm or forest, is undermined by unnatural monoculture or uneven balance, nature hits back and provides the means of eliminating what she does not want. [20]

After the Industrial Revolution, wood had changed from a carefully rationed, essential material to an ordinary commodity, the production of which came to be governed largely by financial considerations, [and] forests were prostituted for profit. Conifers produced the best financial returns, and, in consequence, they were grown more and more, to the exclusion of broad-leaved trees. Certainly there was at first an increase in the financial returns, but the even-aged plantations of a single species were contrary to nature. Before long serious defects were obvious: there was soil deterioration and decreased rate of growth. This system of monoculture lessened the trees' resistance to animal and plant parasites and increased the risk to injuries from snow break and wind. [21]

Of one thing I am convinced: clear-felling is not economic except under very exceptional circumstances. The ideal system is one which will keep the land constantly covered with a forest consisting of uneven-aged trees of various species. What is gained by clear-cutting is too often gained at the expense of the future, for it entails the cutting of many small trees which would have eventually grown into profit. [22]

In my opinion, there should be no clear-cutting at all. Felling should be by selection of the best stems, the mature trees, or by a group selection method where a cluster of trees is removed to enable the surrounding trees to regenerate the land. Planting should be a last resort. Good forestry, good silviculture allows for natural regeneration and planting should only be done in the case of emergency, or on fresh land. [23]

The other thing is that big machinery for felling and extracting the timber must be kept out of the forest. The ticking-over of the engine causes the ground to vibrate, making a hard pan in the soil about ten inches below the surface. The roots of the young trees will not be able to penetrate that hard pan. [24]

Speaking about forestry in Sweden: This great northern forest region remained undisturbed until ninety years ago. Those great stands of pine and spruce had never been touched by man, except in the lower reaches of the rivers; along the valleys were little saw mills. The golden age of the Swedish saw mill industry began when timber tariffs in England and France were lifted. At first these forests suffered grievously, because the plums were taken; the prime timber was removed while the suppressed, dead, and dying trees were left. During the past thirty or forty years, however, public opinion in Sweden has changed all this. Cuttings have been made with due regard to silvicultural principles. Now the overmature trees capable of further growth are retained, while mother seed trees are left to ensure regeneration. Thus the forest capital is maintained to yield its annual wood increment for the wealth of Sweden. [25]

3

Sylvan Heritage

"You can gauge a country's wealth,
its real wealth, by its tree cover."

Sylvan Heritage

More than anything else St. Barbe wanted to awaken people to the rapid destruction of the forest. He sought to convey that the biological and spiritual significance of trees to life on the planet was so much more important than the returns from exploitation of trees for wood, pulp and fuel. He felt we needed to reverse our pattern of taking from the earth without replenishing.

The health and the economic security of the human race depend on how well the forests of the world are managed. All the countries of the world are suffering the penalty resulting from man's neglect to plant where he has reaped. He has destroyed the gifts of a generous Creator without realizing that they were a trust to be handed on to future generations. The earth's green covering is nature's capital, and, if man exacts more than the interest or annual increment, he is endangering the source of wealth and the very means of his existence on the planet. [26]

Man, in his dire folly, has removed vast areas of virgin forest, and now the end of earth's sylvan wealth is in sight. There is no time to be lost if man is to be saved from bringing disaster on this planet and himself. [27]

No one man is capable of doing all the work nature demands to recompense her for past exploitation and neglect. A hundred thousand men could not recompense her. It is the task of Everyman and he must face it and in application learn to appreciate the philosophy of well-being that inspires it.

The planting of trees and reforestation is one of the most vital tasks of the world today. We know how indispensable trees are to the commercial welfare of our nation and how they affect our life, health, climate, soil, rainfall and streams. We would not be without the beauty of the trees in the countryside.

It is only now that the significance of trees is beginning to be understood. Of course trees have always been valued for their timber and, to some extent, for their beauty, but the average person has been blind to the vital functions of the living tree. Yet only in a vague sort of way can we assess the real contribution that trees make to man's existence. Their function is legion and their life is interwoven with that of man and animal, earth and air, food and water. [28]

I think that we should give first priority to saving the breath of life — oxygen. We should all be fighting to save the Amazon forests and to develop a scheme whereby those presently in control of the Amazon can be compensated by the rest of the world for the forests they leave intact. [29]

It is devoutly to be wished that the comparatively extensive tree cover of South America will be preserved, for it is a unique possession in our world of ruthless exploitation of woodland wealth and should be maintained in the interest of the human race.[30]

You can gauge a country's wealth, its real wealth, by its tree cover. In spite of our beautiful parks, Britain is only 6.5 percent wooded, whilst France is 26 percent wooded, Germany 28 percent and Sweden 57 percent. We are almost at the bottom of the list: there is only one country worse than ourselves and that is Ireland. A country's very poor that doesn't have trees. Look at the Sahara: the desert is spreading along a two thousand mile front, in some cases to a depth of thirty miles in one year. It is becoming poverty-stricken. People who have lived for generations on what the forest yields are now having to cut down the forest to make way for cash crops, forced to retreat before the oncoming desert. [31]

The world is slow to recognize the need for great forests close to large centers of population. Legislation seems powerless to guarantee protection or ensure the future existence of the human race. It would seem that nothing short of a universal spiritual regeneration will suffice to change the heart of man and enable him to recognize the law of return and the reasonable demands made by the Creator. [32]

Are we so deafened by the lure of transitory gain that we cannot hear the spirit of the trees appealing to us? Are we so enslaved to the marketplace as to pledge our souls and those who come after us to what we now consider as well-being? [33]

What shall it profit us, as a nation or a dominion, if we balance our budget at the cost of the destruction of the earth beneath our feet? When we find that Kingdom which is within each one of us, and live in it, then all the things we need will be given us. Then and only then, shall we become as wise as the trees. [34]

We must reforest high ground. Not only here in this country [England] with its gentle slopes, but all over the world where bare mountains and lack of water go hand in hand. We must find trees that will flourish above the old tree line, that will grow higher and be more stubborn in the face of nature than any tree we have known before.

For the deserts do not wait, weighing arguments. Deserts are more than scars on the surface of the earth. They are a threat to our existence. They are not static, they do not stay in one place. They are moving over the surface of our arable land at an alarming rate. [35]

Many people will remember the dreadful affair of the felling of the elms in the Broad Walk, Kensington. These fine trees had stood for many years and were good for another half century or longer. Then without warning officialdom decided that they should be removed. Not all the protests of the nation could stop the axes of officialdom. The great elms came down.

Not one of them was suffering from elm disease, not even the four that nearly came down half a century earlier. I took a timber merchant along with me to inspect the huge trunks as they lay on the ground. They were in perfect condition. It was a pity, he said, that they could not have been left for another fifty to a hundred years, when their value would have been much greater.

A tree committee was formed after the catastrophe of the Broad Walk elms. It could not save those trees, but it did achieve one thing; it saved the 4,000 trees in Hyde Park which had also been scheduled for felling. The visitor to London who now finds rest and shade under the trees in the Park can imagine what it would have been like had the axes been unsheathed. Instead of trees, bare gravel paths. Probably, in due course, the grass itself would have been cleared away as untidy, and neatly swept asphalt might have taken its place. We must always be alert. [36]

Perhaps we have forgotten much that is essential to life in our mechanistic age and before we can make peace with the earth we shall have to retrace our steps a long way back and find where we went wrong.... Let us visualize life — all life — as being the outcome of two basic factors: spirit, the invisible cause of all effect and matter, and the outward manifestation of the spirit within. Is it not clear that the same invisible spirit is within the form of a tree as within the form of a man? [37]

[The forest's] uses are numberless and the demands that are made upon it by mankind are numberless also. It is essential for the well-being of mankind that their demands be met, but the length of time required for the growth of a forest shows conclusively that it was never destined in the order of nature for the exclusive use of a single generation. [38]

We're all to blame, for we've all been greedy. We've all been making big demands. We all like to be put into big buildings and houses. We've been cutting and developing recklessly, without a clear definition as to what land should be left for farming, what land should be left for forests and what land should be developed for building. We've gone crazy about making concrete jungles, and we should be grateful to the people who live in the country and who grow trees. [39]

They're teaching new biology in the schools... about the *pyramid of life.* There is the ground producing all the soil bacteria, which is in the top few inches. That grows the grass, and a lamb comes along and eats ten pounds of grass, and that makes one pound of lamb, and then a tiger comes along and eats ten pounds of lamb, and that makes one pound of tiger. We have too many tigers. The pyramid of life is upset, and one of the things we must do is to turn from an animal economy to a silvan economy. We've got to have tree crops, instead of wasting all this land for raising beef. It takes eighteen times more land to feed people on beef than it does on nuts and fruit. Eighteen times more land! When half the human family today are dying from starvation, I don't feel justified in making these demands on the earth. [40]

In some countries... up to three-quarters of the land has been degraded to the use of growing crops to feed animals which they kill to feed themselves. Surely a round-about way of getting food, when it is possible to get food for ourselves direct from the earth through fresh vegetables, fruit and nut-bearing trees. [41]

We are inclined to flatter ourselves that we can control nature. It is thought that science and chemistry, with technical knowledge, have advanced to such a degree that they will solve all our problems. It therefore comes as rather a shock to us that the civilizations of the New World, which have developed with such rapidity, may decline with like speed. Nature is still our ruler and if we attempt to outwit or restrain her, or to transgress her laws, her answer is sterile fields and empty granaries. [42]

Man the destroyer must turn from his mad career and become man the conservator. What still remains of earth's finest forests must be devoutly guarded as a sacred heritage, and as a goal, an example by which to measure the endeavour of future generations.

The trees must be protected for the sake of themselves and the existence of man and not merely to pander to a holiday public. [43]

Trees are needed in the world today as never before. The tremendous material strides made by our present civilization during the last few decades have been largely responsible for bringing about a shortage of wood. The virgin forests of the world are no longer adequate to supply the ever-increasing demands made upon them. [44]

We've got to turn now from this destructive economy to a creative economy — one based on an ethical approach to the earth. Unless we play fair to the earth, we cannot exist physically on this planet.⁴⁵

This brings to mind the teachings of... Bahá'u'lláh [*the Founder of the Bahá'í Faith*], on the indebtedness of man to the earth, his dependence on it and his duty to it. The principle of world economy enunciated by Bahá'u'lláh, begins with the farmer and the care of the land, as this is the most important work. Grave warning of the dangers of shortsighted and selfish use of land are seen in such statements as "My earth is weary of you." ⁴⁶

4

Trees and Life

"Men and trees, water and trees, man and water are inseparable. This is the trinity of life."

Trees and Life

Richard St. Barbe Baker used words like green canopy *and* sponge *to describe the earth's covering. He believed that trees and water were inextricably linked. Without one, the other could not exist. He explored this relationship in his writings throughout his lifetime, while also trying to convey the key role of the forest canopy in agriculture.*

As St. Barbe saw it, the disruption of this water/tree cycle was the first step in the process of bringing on the desert.

I learned early to regard the forest as a society of living things, the greatest of which is the tree. Its *[the tree's]* value depends upon its permanence, its capacity to renew itself, to store water; its many biological functions including that of providing nature's most valuable ground cover, and building up to a great height stores of one of the most adaptable of raw materials: wood. [47]

The truth is, we know all too little about the vital functions of trees. They consume little from the earth and, indeed, give back much more than they take, and as for water, it would seem that they create it. [48]

Trees above all are the beings which attract the waters of the firmament, conserve them in their shade, govern the whole vegetable kingdom in its great economy of water, leading it gently into springs, streams and rivers and maintaining fertile potency in the soil of a region. [49]

Let us see what happens to a droplet of water which has its birth, say, in one of the great oceans. The sun has lifted it from the sea and it has gone up in the atmosphere and perhaps been driven by the winds over the land surface at great heights. It might go on indefinitely, but for the fact that a forest is below and this forest of trees is transpiring moisture into the air. The transpired moisture rises until it meets that single droplet and all the millions upon millions of other droplets. They are reinforced and made heavier by this tree-transpired moisture. A single tree has been recorded to have given off some hundreds of gallons of water every twenty-four hours and a forest of trees creates clouds of moisture above them.

You have probably sat on a hillside [when] the air has been clear; you look out over a valley of woods with a breeze coming towards you from the sea. Cumulus clouds begin to form over the forest — you see them building up as you watch — and later they may come down as rain. If there was no transpired moisture from the land surface, the water droplet with its millions of companions that had been lifted from the sea would have gone on indefinitely until it reached the ocean again, where moisture was still coming up. It might then have come down in the form of rain over the sea.

Trees are the essential link we want to bring back rain to the land.

The roots of trees tap subterranean supplies of water and bring them up to the surface in the stream flow to the leaves. The leaves having fulfulled their function of transpiration and wood formation, fall to the ground.

As we all know, when leaves fall they keep the ground cool. This and the shade of the trees maintain a normal ground temperature.

If the ground is bared and the canopy of trees is opened up

too much, the sun gets into the surface of the forest floor, the temperature is raised and the water is dissipated. The water cycle is broken and growth is reduced.

There is not only a vertical movement of water from the forest through the transpired moisture but there is a horizontal movement as well. Whenever there is a clearing in a forest, the moist air comes out of the forest from all sides on to the clearing and condenses in the form of dew. This is the dew pond principle and in the morning there may be as much as the equivalent of a quarter of an inch of rain which has been condensed onto this patch, which during the night was cooler than the surrounding ground area.

There is another kind of horizontal precipitation such as that along the coasts on a hot summer's day in places like California in the realm of the redwoods. Within ten miles of the Pacific Coast, heavy mists come inland and the tall trees — three hundred feet high — trap the coast mists and form what is called horizontal precipitation. This mist condenses on the leaves of the trees and falls down just as if it were a gentle rain. There is a continual drip from the higher branches and that is how these coast redwoods get their moisture. The hotter the summer, the more water they get through horizontally-precipitated coast mists. Thus it is when trees need more water in the heat of a dry summer, nature provides it.

What happens in a forest when there is a rainstorm? First of all, the fall is broken by the leaf surface of the trees and the water comes gently down to the ground which is protected firstly by a layer of leaves or needles and then by the humus which has been formed by decayed leaves — a sort of compost. Next the water is held up by millions of hair roots of the trees — the feeder roots. The intricate network of support roots and channels formed by old roots that have died in the ground, also trap the water. It [the water] percolates slowly down and forms springs which may come up later, perhaps in the dry season even months afterwards.

When rain falls upon a bare hillside the water runs off, finding the quickest way to the rivers or the sea, and in doing so it probably takes with it valuable soil covering from the hill. So the trees not only conserve water but they also conserve the soil and in this way tend to prevent floods and droughts. [50]

The beneficial influences of a forest are many.... During the day the ground under the trees is protected from the sun's rays and is therefore cooler than soil unprotected. As a result of this protection, the forest air is cooler than the air in the open, and, as air is in constant circulation, this tends to reduce the temperature on hot days. At night, trees retard the radiation of heat from the ground under them.

This helps to equalize the temperature. To anyone flying over a forest in the daytime the upward motion of hot air which has been driven into the forest from the surrounding plains is quite perceptible, while at night there is a reverse process which tends to drag the plane down over any extensive forest area.

The water capital of the earth may be thought of as consisting of two parts, the fixed capital and the circulating capital. The first consists not only of the water stored in the earth, but also of that in the atmosphere. The circulating capital is that which is evaporated from water surfaces, from the soil and from vegetation, and which, after having been temporarily held in the atmosphere in the form of clouds, is returned again as rain, dew, or snow. The function of the trees is to filter the water and distribute it, and forest cover directly influences the pattern of this distribution.

We cannot yet clearly follow the intricacies of the circulation of water on the earth, in the earth, in the trees, in the ocean, and in the air, but we do know that man's ignorant interference with the forest and other cover has greatly reduced the earth's fertility and is still doing so. The natural circulation of water is a subject which must have our careful study and deepest consideration. [51]

Men and trees, water and trees, man and water are inseparable. This is the trinity of life. A chemical analysis of the human body shows that it is ninety percent water. Similarly, a tree is found to be ninety percent water: and yet blood is not water any more than sap is water. If you try to inject, say, a quart of water into the bloodstream of a man, he would die; similarly if you inject water into a tree, or even place its roots in water for any length of time, it will die — and yet it has been proved that a high tree will transpire as much as from 100 gallons to 500 gallons of water a day into the air. Where does that water come from? [52]

It *[the tree]* is a miracle of growth from seed to maturity. From the kindly darkness of the earth, where germination takes place, the shoot and root progress in opposite directions, the former drawn towards the light, the latter digging into mother earth for sustenance. The root explores for moisture, and sends out little branch roots in search of food, which stimulates the upward growth. The shoot in its turn sends out branches, and through the activity of the cambium, or growing layer, the stem thickens. Supplied with a few mineral salts, water and air, the tree builds up its body, increasing in size as year by year it adds layers of tissue to its growth. The essential chemical processes continue under the influence of sunlight, and by reason of its unique power of utilizing radiant energy from the sun, the tree obtains carbon for the manufacture of carbohydrates and proteins.

The green leaves which clothe the tree may be likened to the laboratories where carbohydrates and proteins are made, for all the processes of growth take place in them. They build up the framework and add fuel for the maintenance of respiration and the working activities of the whole tree; thus life and growth depend upon the leaves.

Another miracle is the manner by which trees cable themselves to the earth and maintain their poise while continually tapping subterranean supplies of water. The roots of a tree serve a double purpose, for while anchoring the tree and providing a mechanical service, they actively take in water, though it is only the younger and more delicate parts of the root system which do this.

The root tips are provided with remarkable... apparatus for the purpose of taking up water and mineral food from the soil.

The water, drawn from the depths of the earth, continues to rise by osmotic pressure unchecked... until it reaches the topmost leaves of the tree. [53]

A man who plants a tree is doing a very wonderful thing. He is setting in motion an organism which may far outlive him or his children, and year by year that tree is storing up energy and power, working with precision like a factory, but far superior to any factory of man. On a full grown tree there are thousands and thousands of leaves constantly working, evaporating and breathing out water into the atmosphere. Think of the enormous surface of these leaves. Take for instance a large elm in one of our London parks. If you were to defoliate it and spread all the leaves side by side on the ground, the leaves from a single tree would cover ten acres of land. Now imagine those leaves covered with minute stomata, each working, absorbing the maximum quantity of solar energy, nitrogen [from] the atmosphere and various other elements and minerals, and many different radiations which they accumulate and elaborate by means of millions and millions of chlorophyll cells.

Besides water, trees provide pure air. They are the great filtering machines for the human organism. They improve and transform the air in a way which is most favourable and most acceptable to the lungs of man. [54]

According to ancient mythology, trees were the first living things on earth. This is borne out by scientific reasoning which shows that it is through them that the air we breathe can give life to humanity. Through countless ages trees have been drawing carbonic acid gas from the atmosphere, absorbing and incorporating the carbon, assimilating it; then when they die, bequeathing to soil their carboniferous remains. The consequence has been that eventually the atmospheric oxygen was left sufficiently pure for the requirements of birds and mammals which have replaced the flying reptiles and monstrous amphibians that were able to endure the heavy air of primeval swamps and jungles. [55]

In the south of England we have a good example of the part which trees play in the water cycle. Up on the downs above the racecourse at Goodwood there is a very fine forest of beech. Part of it was clear-felled. What happened? The water table sank, the springs which were formerly near the surface sank in the ground and water quickly found its way to the sea. [56]

Water must be a basic consideration in everything: forestry, agriculture and industry. The forest is the mother of the rivers. First we must restore the tree cover to fix the soil, prevent too quick run-off, and steady springs, streams and rivers. We must restore the natural motion of our rivers and, in so doing, we shall restore their vitalizing functions. A river flowing naturally, with its bends, broads, and narrows, has the motion of the blood in our arteries, with its inward rotation, tension and relaxation. Picture a river which has risen from a mountain spring in a well-treed watershed: trees of mixed species and different shaped roots; the spear-shaped roots, heart- shaped roots and flat roots, fixing the soil at different levels and reducing competition for food and water. The leaf fall and humus on the floor of the forest will act as a sponge to retard quick run-off after a storm. Water will sink through to porous soil and form myriads of springs which will feed the land and the rivers during the drier months of the year.

Mountains and high ground should be covered with protective forests up to the snow-line; in high country, fields should be kept small and carved out of the forest and always be tree-surrounded. How strange it is that communities fail to realize the importance of preserving tree cover on tree slopes. [57]

Referring to his foresty practices in Nigeria: In the mahogany forests I saw to it that the soil was never exposed to the sun, rain, or wind. The rain was broken up by the canopy into fine spray; the ground was permeable; the soil below readily drank in the rainfall. I carefully avoided anything in the nature of weeding or unnecessary cultivation. Vast quantities of water from the rainstorm and the river are held up in transit by the thick carpet of permeable humus. In this way my forest acted as a huge reservoir only gradually releasing the water in the form of springs into the clear, deep, slow-flowing Jamieson River. [58]

Thus the tree, with the help of all plant life, controls the food supply and life of man and of the animal kingdom. [59]

The productive capacity of the soil depends on trees, and agriculture is dependent on forestry: the one devoted to the planting, conservation, and harvesting of trees, and the other to grains, fruits, vegetables, and animals.

Every farm contains land that will not produce farm crops, but no land is of such poor quality that it will not support tree growth. If the farmer will only recognize the value of the trees to himself and the importance of the forest to the country and to mankind in general, he will undertake to grow crops of trees with as much care and forethought as he spends on producing other crops. Forestry and agriculture are complementary and may be practiced on the same farm. The forest provides fuel and construction timber, fence posts and wood for implements, and, while growing, the trees provide shelter for stock and nesting places for helpful birds that destroy harmful insects.

Above all, forests increase the fertility of the land. It is possible, by planting unproductive areas or the lower grades of farmland, to improve their quality. With the exception of steep hillsides or sand, all land can be brought within the range of agriculture through the improvement of the soil made possible by the leaf fall and irrigation of forest trees. It is therefore evident that the economy of forestry and that of agriculture are intimately connected. So close is this integration that when a farmer suddenly sells his entire wood crop and removes his hedgerow timber, a severe threat is made upon the economy of the farm. To fell all the trees on a property is often a prelude to disaster. In many parts of the world, farm abandonment and decline of agriculture have coincided with forest destruction. [60]

Trees create micro-climates, reduce the speed of wind, lift the water table and increase the population of worms.... If farmers only knew how to harness worms, they could double their crops. Trees provide the answer.... If you want to double your supplies of food, then you should devote twenty-two percent of your farm to trees, to strategically-planted shelter belts. [61]

Upon crossing into Nigeria on his Sahara expedition: On we went through desert or semi-arid waste, of savannah and thorn scrub country. Abruptly we reached a watered region along the boundary of Nigeria. What a contrast! In two hours we had passed from an arid waste to a rich agricultural region where groundnuts were the staple crop. They were not grown in great fields but among large forest trees, in little clearings in the high tropic forest. Now this was very interesting, for they were grown in small patches, tree-surrounded. At nighttime their leaves would fold up like butterflies' wings at rest. When the hot air from the surrounding forest passes over the cooler patch of the groundnuts, it condenses and in the morning there is a heavy dew equivalent to as much as a quarter of an inch of rain. As the sun rises the leaves of the groundnut plants open out horizontally and again cover the ground with their shade, keeping it cool. When there is heavy rain it tends to wash the soil away from this patch; but it does not go far, because the root system and humus of surrounding trees hold it.

This is the secret of the successful growing of groundnuts, which had been known to the Nigerians for many generations. [By] preference the patch is sufficiently small so that for the greater part of the day it is shaded by the surrounding trees. Down through the ages the Nigerian peasant has evolved a silviculture, an indigenous creative economy; instinctively he knows and appreciates the vital functions of the friendly trees, which enable him to produce rich crops and at the same time protect his land from the threat of the desert. [62]

Dr. Halliday Sutherland records that since pioneering days "down under", the water table, as ascertained in the sinking of wells, has continuously fallen with the destruction of trees to make room for sheep and other farming. At one time there was a veritable freshwater subterranean sea under much of Australia. The trees there sent down their roots to almost incredible depths through the subsoil to reach it and could not otherwise have lived through long droughts. Now, even when not directly destroyed by man, they die because the water table has sunk farther than they can reach.[63]

5

Skin of the Earth

"I reverently held that teeming soil
before letting it filter through my fingers
back to the earth."

The Skin of the Earth

For St. Barbe it was clear that the health of the soil was dependent upon tree cover. Once deforested and exposed to the sun, wind and rain, the skin of the earth was no longer able to defend itself and maintain its fragile equilibrium. The rich humus essential to the support of vegetation, and thus all life on the planet, was easily dried up, blown away or swept away in a rainstorm.

We can... understand why forests are the ideal covering for the earth, for they maintain the right soil conditions for plant growth, the litter of leaves, twigs, and branches shed by the trees which completely cover the ground. This litter may be only two inches in depth, or it may be a foot or more, but the effect is the same. Its first function is to keep the soil from being washed away, but second only to this in importance is that it retains moisture and rain water, which instead of running off, seeps slowly into the earth, to come up later in the form of springs. This economy preserves the soil in the necessary condition to fulfill its major purpose of producing plant and food growth. [64]

As long as soil is covered with forest, its humus is maintained. The basic forest problem lies in its composition and regeneration. In the forest, the processes of decay and growth always balance one another. The vegetable wastes together with the by-products of the animal population form a mixture on the forest floor. As we examine this mixture from time to time we find it remains practically constant in depth, in spite of annual additions from leaf fall that take place. This mixture is drawn upon at an even rate by earthworms, fungi and bacteria, and the resulting humus in turn is absorbed by the soil and provides the trees and undergrowth with the food materials they require. Thus the forest manures itself and, with the help of the earthworms and other animals, distributes this manure through the upper layers of the soil. Everything is done by nature quietly and efficiently. No artificial fertilizers, no selective weed-killers, no pesticides and no machinery are needed in the household of the natural forest. [65]

The trees and vegetation, which cover the land surface of the earth and delight the eye, are performing vital tasks incumbent upon the vegetable world in nature. The glorious rich, colorful, quilted covering of vegetation is not there merely to feed and please us, its presence is essential to earth as an organism. It is the first condition of all life; it is the *skin* of the earth.

The vegetation... has the same functions as the skin on the human or the animal body. [66]

If a man loses one-third of his skin he dies; the plastic surgeons say, "He's had it." If a tree loses one-third of its bark it dies.... Would it not be reasonable to suggest that if the earth loses more than a third of its green mantle and tree cover, it will assuredly die? The water table will sink beyond recall and life will become impossible.[67]

Vegetation constitutes the foundation of our energy and the very source of all life, for in the final analysis this applies even to carnivorous animals. The only difference is that the carnivores obtain the natural energies stored in plants secondhand, while the vegetarian animals receive them directly. Man also... draws his vitality from the same vegetable world, which in turn springs from the soil. [68]

When the Australian farmer uses the term *planting*, he usually means sowing grass seed or farm crops. This is done after the felled forest has been burned off.... Almost anything will grow without cultivation in this ash from the trees that have protected and enriched the soil for a thousand years; though these methods are nothing less than despoiling the earth of the stored-up wealth of the ages. [69]

Forest destruction in every country spells disaster in one form or another. Whereas, for instance, in Africa it brings erosion and deserts and in Switzerland avalanches, in Italy the rivers are turned into torrents and floods by deforestation. [70]

In the course of three centuries Britain has colonized large parts of the earth's surface. We *opened up* these lands — literally *opened them up* to the glare of the sun. In our own gentle climate it would not have mattered, but in Africa, Australia, or New Zealand the balance of nature was destroyed and the first beginnings of a desert were made. Trapping and holding rain is as important as the conditions that produce it. Clearing the trees and opening the land to the glare of the sun does not always mean less rainfall but it usually means that what rainfall there is quickly evaporates and runs off the land. Often it takes the rich topsoil with it. [71]

Resulting from unnatural industries, from war, and from greed, the modern world has reached such a degree of mechanism that the soil of one part of the earth is transported to another to feed those who make no contribution themselves to its fertility. In the early days of World War II, people were shocked at the reports that the Germans were sending trainloads of good black earth from Czechoslovakian forests each day into Germany to reinforce the worked-out lands of their own country. But little does the public realize that this process has been going on throughout the world for the past century. For instance, when the early settlers found themselves in the newly discovered continent of Australia, they took to sheep farming and wheat growing because they could make big money from these. Their flocks grazed on virgin lands and they felled the forest to extend their farms. They exported their wheat and wool harvests to other lands and with them their land's fertility. As the market boomed, their flocks and farms increased and soon the sparse covering of grass was over-grazed and the thin layer of topsoil worn away, exported in the form of wheat, wool, and mutton.

The taking of food from the topsoil and failing to return it in compost is suicidal. Western man has been undermining his very existence since the introduction of chemical manures and water sanitation. He has been raping the earth, taking from it crops of every kind, and has failed to recognize the law of return and fair play. Few people realize that every ton of wheat grain represents four fifths of a ton of earth.... In so far as this produce is fed to livestock, it may be returned to the land as farmyard manure; but the great majority of food consumed by city dwellers eventually leaves their bodies in the form of solid and liquid waste and is lost to the land, owing to modern sewage. [72]

In the autumn of 1910, while crossing the prairies of Canada, I recognized for the first time a desert in the making. Wide areas had been ploughed up where for centuries dwarf willows had stabilized the deep, rich, black soil. The country had been divided into townships with sections of 640 acres. In those days anybody could file on to a quarter section of 160 acres for nothing, and if he needed more, that could be acquired.

The first thing they did was to plough as much of it as they could, then sow wheat and oats to feed the horses. One could travel miles without seeing a tree. When a farmer took up a section of land, he would mark the boundary, put up a couple of poles with strips of an old white shirt tied on the tops, so that they could be seen a mile away; sit on the plough with his six-horse outfit and drive straight, keeping the markers between the heads of the leading horses — backwards and forwards on dead level ground, breaking five acres a day. Two crops of wheat would be grown and then a crop of oats. His neighbor would be doing the same. With no sheltering trees, the soil began to drift and blow away; up to an inch of soil would be lost in a year. [73]

Where forests disappear the desert soon takes their place. Trees hold the rain as it falls and condense the fogs, precipitating their moisture. When the trees are gone, rainwater rushes down the hollows and valleys, cutting deep water courses which carry off the water before it can saturate the ground. The mists, no longer held in the foliage, drift away without depositing moisture. The rushing stream carries in its flood the soil and fertile humus. The springs are not fed, because rain has not time to sink down to their level. Streams and rills become torrents in the wet season, and barren ravines in the dry. [74]

Whenever tree cover has been removed or interfered with by grazing... sheet erosion takes place. Tiny rivulets carry away the soil in small quantities at first, but as the gullies widen and concentrate the water, it soon carries away bigger loads until a single storm may account for the removal of thousands of tons of good earth. This is an ever-present threat to the farmer. [75]

Following the removal of the high forest with its dense ground covering, the rainfall tends to become less evenly distributed. When a storm breaks, run-off quickly brings erosion to the land. Gullies are among the more spectacular results of erosion which play havoc with hilly land, while rolling land is rapidly ruined by the more insidious and widely destructive process of sheet erosion. It has been demonstrated that when gully formation has begun, the land has already lost much of the most important element in soil fertility and stability — namely, the water-holding capacity of the soil. This is apart from the actual soil material which has been lost by wind and rain. In Australia, generally speaking, the water-holding capacity of the soil is confined mainly to a few inches on the surface, where fresh humus formed from decaying leaves and animal remains accumulate. Sheet erosion, by removing the most absorbent layers, not only speeds up the run-off water which is the chief eroding agent, but decreases the value and shortens the period of usefulness of the rainfall. In a country where every drop of rain is needed to maintain life, this factor of erosion is more serious than the actual loss of soil. [76]

Describing erosion in China: The good earth... must be tended with loving care; otherwise it will not support [man].... Her skin is sensitive to the behavior of man, who has never quite mastered the art of continuous cultivation of land which was once covered with forest. The folly of exposing her nakedness has been shown graphically in China. Deserts now reign in the very birthplaces of civilization. The great northwest loessial region now resembles a huge battlefield, scarred and denuded. That fertile soil which, when touched by rain, will bear crops two or three times a year easily dissipates itself unless covered by trees or grasses.

The name "loess" is adopted from that of a Tertiary deposit which appears in the Rhine Valley. It is a brownish-colored earth, extremely porous; it crumbles easily between the fingers, and it readily forms clouds of dust. The wind-blown loess is one of the richest soil materials known and it is extremely sensitive to erosion which will cut it into deep clefts, making traveling across its surface impossible. Over vast areas, where once the soil was deep and fertile, little remains but gaping chasms, sometimes hundreds of feet wide and deep. The eroded material has been spread on the valleys and plains, and has found its way to the rivers and sea. The Yellow River and the Yellow Sea are named from the color of the eroded soil which has been lost to the hinterland.

9 St. Barbe assisted with the reforestation component of the Civilian Conservation Corps (CCC) in the 1930's.

10 Reforestation crew planting in a burned-over area.

11 St. Barbe led the Sahara Expedition in 1952, and is shown here in Nairobi after crossing the desert.

12 In London's Trafalgar Square on the eve of departure, people brought peach stones for planting along the expedition's route.

13 The expedition set out to observe how, and at what speed, the desert was encroaching on fertile land.

14 An experiment using trees for dune stabilization in Northern Africa, 1961.

15 Greeting guests attending the World Forestry Charter Gathering in London in 1956.

16 A Men of the Trees exhibition in the 1950's.

It has been estimated that the Yellow River alone transports an annual load of 2,500 million tons of soil. There are other regions where rapid erosion is taking place, but the Yellow River has been described as the outstanding and eternal symbol of the mortality of civilization.[77]

Writing about erosion in Algeria: There are other factors in the African climate which intervene to aggravate the phenomenon of erosion; in particular the intense heat of the soil by day following low night temperatures, which together with a long summer drought, contribute to break up and undercut the soil and rocks and thus bring about the erosive effects produced by the first rainstorms, which mark the end of summer or beginning of autumn. The steep hillsides combined with the lengths of the slopes also help to speed destructive erosion.... When the brutal forces of the torrents have been released from the restraint which a permanent cover of vegetation provides, the soil is swept away. Before this natural covering is removed, nature can defend itself and maintain a certain equilibrium; but man has intervened and disturbed this.

Prehistoric man made little impression, but later the depredations of domestic animals and successive marginal cultivations caused land to become barren as the needs of the population increased.... The destruction of forests continued without interruption and is all the more serious because each successive invasion has sent more people to the mountains. What pioneers have mistakenly referred to as an improvement in the land has in reality consisted of a cruel exploitation of the vegetable treasures of the earth which had been patiently accumulated by nature during thousands of years.

It is strange that the cultivator has invariably suffered from the influence of straight lines. He is hypnotized by rectilinear culture, without any consideration for relief or gradient. Thus he prepares converging paths where rushing waters come together. [78]

An iron plough is a dangerous implement in Equatorial Africa because it loosens the earth to a considerable depth, allowing the soil to be washed away in the first torrential downpour. Their *[the native people's]* salvation lies in keeping to small fields, tree surrounded. So long as they only grow food for themselves will they be safe. It is only when they are tempted to clear extensive areas for the production of cash crops such as cotton, that the balance of nature is threatened and, with it, their existence. [79]

Reckoned in forest destroyed, cotton is the most exacting crop and is making greater demands on the earth than any other. By its means the forests of the Congo are disappearing faster than ever before. Another cause of progressive dessication is the annual firing of savannah and grasslands to provide fresh grazing for cattle. Each time it is fired, any accumulation of humus is burned and the quality of the grass deteriorates, so that in a few years' time an inferior species takes the place of what was once a nutritious crop. [80]

Writing about erosion and land reclamation in New Zealand: A previous expedition had taken me to North Otago where farmers had admitted the fact that they had lost the top three or four inches of the soil since they had begun to plough their land, and that each year they were being forced to plough less deep because the topsoil had been blown away. Such was the plight of most of the farmers of the region who had depended on artificial fertilizers for their crops. The earth was being eaten away, its fertility exported, its millions of economic biologic soil bacteria being stifled, poisoned by overdoses of commercial chemical compounds.

Year by year the situation had deteriorated, with tragic results. The murky rivers overflowing their banks told the sad story of deforestation and land erosion. I had flown over the South Island and had been shocked to see the signs of disintegration, the widespread flooding of the plains and the washing away of good soil by the unleashed waters that followed a quick thaw or extra heavy downpour. As we flew along the coast, I could see the telltale signs of discoloured sea for miles along the coast or right out to sea wherever a river discharged itself into the ocean.

How happy I was to find, at Palmerston, New Zealand, one who had applied his practical knowledge to the healing of the earth with the help of compost, worms and the myriads of soil bacteria that flourished in his spacious, happy fields. Mr. Trevor Ross told me how, when he started, his eminent neighbors had ridiculed his ideas and even regarded him as a simpleton, but he had persisted, and soon his enthusiasm and example became infectious, so that now there is a growing organic movement among his neighbors.

How glad he was to welcome me and show me his flourishing farm where he had, by organic methods, increased the depth of his soil over wide fields to an extent of six inches. He would stop in a newly ploughed field and scoop up a handful of earth with the reverence of a communicant receiving the Holy Eucharist, and he would inhale the fresh scent, like that of a flower or a fruit, and look deeply and searchingly into this microcosm of life — without form, yet containing a myriad forms of life....

With unbounded pride and love, he handed me the earth of his resurrecting for my critical appraisal, expectantly awaiting a look of approbation.... My young farm friend was a true husbandman... for that which I held in my hands was teeming with life. I could see but two little, red, wriggly worms, but I sensed the presence of many other invertebrate creatures and countless protozoa. As I peered into the little world within my hands, I could see a weft of fine threads ramifying between the soil crumbs and particles, but for every streak of silken thread visible to the eye I knew that with my powerful microscope I would recognize thousands of different kinds of fungi... and millions of bacteria. These thoughts were passing through my mind as, for a few moments, I reverently held that teeming soil, before letting it filter through my fingers back to earth. [81]

6

Ancient Groves

"In the stillness of these mighty woods,
man is made aware of the divine."

Ancient Groves

St. Barbe saw in virgin groves, whether in the Congo jungle, the towering kauri forests of New Zealand, or the magnificent redwoods of California, a link with our ancient past, as well as an opportunity to observe the rhythm of nature unmarred by man.

He felt that we should preserve what we had and make no further inroads into these vanishing forests until we learned what they had to teach us.

How can I describe the wonder of these redwoods? Is it their age, or beauty, or perfect poise, which impresses one most? Upon entering those groves a spirit of awe and reverence came over me.... In the stillness of these mighty woods, man is made aware of the divine. [82]

Millions of years ago when the earth was young, trees of gigantic stature and girth towered higher than St. Paul's dome and fought with each other in bloodless conflict for light and space. It was a world of trees, huge trees and strange creatures, reptiles and amphibia, dinosaurs and prehistoric mammals. The atmosphere was still foetid but the giant mosses, the first vegetable life which came as the earth cooled down, were now disappearing and the trees were gradually clearing the air, absorbing the gaseous vapours and beginning their terrific task of creating oxygen vital to life.

For millions of years they *[the giant trees]* reigned supreme and adapted themselves to the mountain slopes or plains to which they migrated. At that time the northern hemisphere was one continuous land mass, for Asia was joined to Alaska by a land bridge. Over the whole of this mysterious realm reigned the *big trees*. Other primitive forms, such as the ginkgo, the maidenhair tree, dominated mountain sides in what is today China and Japan. There were trees of lesser growth, agathas, araucarias, monkey puzzles, and, in New Zealand, the Land of the Shining Coast, were the giant kauri trees. The continent of Australia, whatever its land formation, had primitive types of eucalyptus and acacia; the mountains of India sheltered vast cedar forests, which had migrated to these heights in times of flood, and through countless centuries adapted their physical structure to contend with altering climatic conditions.

Through all this time the *big trees* survived, showing little change in form or structure as may be seen by examining and comparing the fossilized remains of forty-five of their species which have been described from the fossil beds of the Eocene and Miocene from the northern hemisphere, including some twelve species from North America. They are in France, Switzerland, Austria, Bohemia, Germany, England and Spitzbergen. In North America they grew at the mouth of the Mackenzie River, in Alaska, in various parts of the Rocky Mountains, in Oregon and in California. We may suppose that during the glacial epochs in Europe they were driven down to the Mediterranean and there perished, since they could move no further southward. In North America the climatic factors were such as to favor their persistence only in California. These redwoods then are in many ways the most sublime type of earth's forests. For thousands of years they have been growing here — the finest example we have on the earth today of the vegetation of the Miocene epoch. [83]

Along the coast of the blue Pacific, from Monterey... to the Oregon border, stretches the Redwood Empire. It is the realm of the tallest and most ancient living trees, and to me its beauty is unrivalled. Throughout this woodland region, there is an atmosphere of calm which gives one an all-pervading sense of mystery and enchantment. Unreal! A fairy world! [84]

It was to the future of the redwoods that I dedicated myself in those days of depression nearly a quarter of a century ago. To many the 1930's were a decade of despair but the creative drive of President Roosevelt's forestry plans opened up a vast sector of the Green Front [*St. Barbe Baker's campaign for land reclamation*] in the United States. In that sector of the peaceful and fruitful battle for the preservation of forests I spent the years that ended in 1939. The struggle filled my days and the faith and love I had for it filled my heart. [85]

I was coming into a more open park-like land and could see across a grassy meadow. For the first time, the tops of the *big trees* were visible. The ground was soft and damp and where I came to the first opening was a massive tree, lying its length of close on three hundred feet. I wondered if it had been felled by fire or tornado. It was imbedded deep in the earth and I rode along its enormous trunk, with its soft bark still intact after maybe a thousand prostrate years.

Here was the forest climax. The underlying reason for the longevity of this forest species is that in [this] region there has been perfect balance between sequoias and surrounding conditions. The requirements of each tree were met: depth and composition of soil, humidity and warmth.

Was there any limit to their growth? Why were they not even greater still? Why were trees which might be four thousand years but a little larger and no taller than those of fifteen hundred years? Why did they not luxuriate and become even mightier yet? The answer to this question may be found upon examination of the upturned roots of a fallen tree. Great as the roots may be, they are moderate by comparison with the tree's size. If they were not restricted they might rampage over immense areas exhausting the soil too far removed from the mother tree to be replenished by its compensating leaf fall. So here again we may observe the principle of balance. Even the greatest things of the world have their weaknesses, while the humblest creatures have their means of protection. [86]

Unlike *Sequoia gigantea* which only reproduces itself from seed, *Sequoia sempervirens,* the ever living, contrives to reproduce its species from dozens of stool shoots thrown up around the stump of a parent tree. These contest with each other for light and air, grow and flourish for a thousand or perhaps two thousand years, when they in turn, before they perish, throw up young shoots from the parent root which in their turn, again, compete with their brethren, live their span and again give life to another generation of trees that may last another two thousand years. I have seen great circles of columns, two hundred feet or more in diameter which careful measurement show may have sprung from a parent tree which flourished nine thousand years ago. Here is life unending; protected now, their cycle will evolve through the unknown future. No tree can challenge the longevity of their line. How can man do justice or pay tribute to living things that reach towards immortality? [87]

Shall we ever probe their secrets? In their presence man is overpowered by a sense of his insignificance, for the redwoods are the supreme achievement of tree growth in the world today, having taken million of years to attain their stature and perfection. They are indeed the temple of the Almighty, and yet man, who should have been their protector, was their destroyer. [88]

Here, in the groves of these immortals, I wondered whether I had been witnessing the *forest climatic climax* of the text-books or whether these forests were still in the state of becoming. There was always the broader question, did the growth depend upon the climate or the climate upon the species which composed the tree cover?

Surely nature could retain here the prerequistite condition for superlative growth by maintaining the compostition it had developed through the ages. [89]

The entire redwood belt originally contained approximately one million acres, more than one-third of which has now been cut over [*figures from the early 1940's*]. The narrow fringe of these monarchs along the Redwood Empire highway must be backed up by considerable areas of forest to support them, for if the lumbermen get down to this fringe, the trees will be exposed to the full blast of the wind and, in consequence, will blow over and be destroyed. Then there is the risk of the climate being changed as the result of the destruction of great forest areas.

Who can deny that the superb climate of California owes much to the effect of the redwoods, if it is not entirely enriched by their influence? No one as yet can tell the extent of this influence, but my personal feeling is that it is very considerable.... The redwoods condense and filter out drops of water from the coast mists, which are often heaviest in the driest summers. This so-called horizontal precipitation may be of considerable benefit to the growth of plant life... and turn what might be a barren land without them, into a state of fertility. For that reason alone, very extensive areas of these healthy life-giving trees should be preserved, even apart from the future welfare of the trees themselves. [90]

I was alarmed at the rate at which these lovely giants were being cleared for building material.

The problem was to preserve... a sufficiently large area... for the balance of nature to remain undisturbed. The trees had to be allowed to flourish and reproduce themselves in the atmosphere they had created. [91]

The struggle to save the redwoods still goes on; it seems that each generation will have to fight to maintain its redwood heritage. [92]

Here man may acquire a nearer perspective of himself in relation to the trees. Here he may regain a sense of the rhythm of nature and be attuned to the universe. [93]

Describing a tropical forest: For years we had known of the great Congo Forest. Here we were near the heart of it, at Aturi, in the middle of what we had heard of as being the last great tropical forest in Africa.

Here were the family of *Leguminosae* and trees of the *Meliatious* family including the great mahoganies and African walnuts, choice ornamental and aromatic woods. I felt at home for I recognized many old tree friends from the rain forests of Nigeria, and the same woody lianas entangling the younger trees with a close canopy shutting out the fierce sunlight.

Through all this ran the wide road, cutting through the dense forest complex and enabling the traveller to view a cross section. It was a superb picture of tree growth in infinite variety; trees of a multitude of species rarely seen in mixture; a paradise for botanists and plant hunters. Orchids were there too, growing high up on the trunks of tall trees, hanging in festoons amid the shade of the leaves among cheeky little parrots, chattering monkeys and rare butterflies. On each side, in light and shadow, tree boles standing free of the usual undergrowth and entangling lianas showed... clearly at the side of the wide road, towering a hundred feet or more, a sight such as is seen nowhere in the world. I was thrilled! I longed for this scene to go on forever.

Alas, suddenly the forest stopped and we came into a great burnt clearing. At first it was unplanted ground, still hot from burning, with great smouldering trunks and piles of wood ash. It was to be a new plantation, part of an enormous paw-paw estate. Each year an additional area is cleared to provide fresh fertile soil to replace that which has been worked out by cropping. [94]

Describing a Nigerian forest: The impression one gets in the forest is that there is ruthless struggle going on all the time — a struggle for survival, the strong suppressing the weak. It is almost impossible to follow what is happening in the undergrowth above the ground so tangled are the creepers and woody lianas. But in this zone lies the future destiny of the tree.

From the ground it is difficult, if not impossible, to get a clear view of the treetops. When I was first stationed in Benin I used to bicycle out before breakfast, first in one direction and then another, hoping to find a point of vantage from which I could study life in the treetops. But at Sapoba, with the help of a forest dweller I made a ladder and climbed to the top of the tallest tree, where we made a comfortable platform from which I could look out on the whole dazzling life of the treetops. It was like being in another world to recognize the crowns of the trees I had known so well on the ground.

How beautiful the whole scene was with tiny birds fluttering from flower to flower, butterflies camouflaged against their feathered enemies! I found myself wondering if the baby mahoganies I had entered on the charts of my quadrats, would ever reach these exalted places. [95]

St. Barbe's experience of kauri forest in New Zealand: I shall never forget my first sight of these lordly trees in the Waipoua Forest. They stood in the virgin bush, which had been opened up by a clearing, so that their mighty buttresses appeared to form a colonnade. Overawed by their kingly dignity, I felt that I might have been standing in an ancient Greek temple, for their huge trunks stood out in clear-cut relief backed by the somber green of lesser forest growth.

A remarkable feature of the kauri is its trunk, which has hardly any taper, but grows a cylindrical bole over one hundred feet high, surmounted by an umbrella-shaped crown of olive-green foliage.

Although the lower parts of the trunks were usually branchless, they did not give the appearance of being bare, for these patriarch trees acted as hosts to a number of climbing ferns, lichens, and hanging grasses, while air plants and orchids draped their massive columns. Tough, woody lianas hung suspended in the air, and bushes, trees and shrubs of every kind formed a tangled mass to cover the ground around their buttresses.

There was an incessant struggle for life, silent and serene. As I entered the forest parthenon, rays of light shot through the foliage and sunbeams danced on the fronds of climbing ferns and pendant grasses, and dappled the gray-brown trunks of those ancient friendly trees. They seemed to be beckoning me on into the inner recesses of a quiet world where time is not. I found myself standing close to the oldest tree of all.... Eleven generations of his family had germinated, grown, and fought their way up through the tangled growth, looked out over the treetops where they reigned for one thousand years, and then returned to earth to complete the cycle before being reincarnated in their grandsons. For one hundred and ten centuries the silence of the forest was unbroken but for the note of the bellbird and his many feathered friends. [96]

As a forest officer in Equatorial Africa, I have had an opportunity of studying the rapid deterioration that is taking place in forest types. I have seen the forest climax where superb trees vie with each other in the battle of terrific growth and where the forest has reached perfection. I have seen this same area invaded by nomadic farmers who have cleared away the growth of ages and sacrificed some of the most valuable timber in the world, burning it to make ash in which to grow their yams. I have marched for days along forest trails through what to the casual observer might have appeared to be primeval forest, but upon closer examination I found that the whole of this area was pock-marked with clearings. The rhythm had been broken — never again would that forest attain its former glory. [97]

7

Desert Challenge

"What appears to be an inexhaustible treasure-laden earth is slowly becoming a plundered planet."

Desert Challenge

St. Barbe outlined how the essential relationship between water, trees and soil becomes disrupted, primarily through agriculture and through the felling of trees for wood and fuel. It is this disruption that inevitably turns land into desert by progressive stages, and it was this concern which became the focus of his life's work.

It may be observed that there are various steps of degradation from forests to desert. When the forest is cleared for farming or other reasons, the debris is sooner or later burnt up. The humus accumulated over thousands of years is thus destroyed and the earth exposed to the elements. Water is dependent on the tree cover. When the trees go, the natural circulation of water is broken, the spring water table sinks and the lift of water resulting from transpiration is relaxed.

It is conceivable that for every hundred litres that a forest tree transpires into the air, a hundred and fifty litres is brought to the surface and sent surging through the surface soil. Whenever an area of more than twenty-five acres is cleared there is a danger of the spring or surface water table sinking beyond recall; how much more dangerous and detrimental would be the wholesale destruction of a forest over many hundreds of square miles. Even when this happens, wind and birds would bring seeds into the cut-over areas and in time would rehabilitate the land. But with the presence of large flocks and herds of rapacious animals, few trees, if any, are allowed to grow again. Trees are essential not only in the maintenance of water circulation; they are truly the skin of the earth. When they are removed the humus disappears and the stored up fertility of a million years may be lost in a single season.

Without tree cover the temperature of the earth rises, and cloud formations pass over without making any contribution to moisture. Moreover such moisture as there may still be will quickly disappear when not protected by shade, leaving soil which, if not washed away in a torrential downpour, will eventually be carried away by the increasing force of the winds. [98]

The unbridled avarice of man is in nearly every continent of the earth destroying the biological balance. What appears to be an inexhaustible treasure-laden earth is slowly but surely becoming a plundered planet. Our woods and forests, the indispensable lungs of our earth-organism, are falling into a murderous dance of death. [99]

Of the earth's thirty billion acres, already more than nine billion acres are desert. Land is being lost to agriculture and forestry much faster than it is being reclaimed. At the same time the world population is exploding. Already half the human family is on the verge of starvation, for man breeds and lives beyond the limits of the land. [100]

There are many reasons advanced as to the cause of the Sahara but most of these omit the stages from forest to desert. To me, the conditions in the Sahara clearly prove these stages and I do not believe that any sudden cataclysm of natural land upheaval was responsible.

When a forest is felled for farming or other reasons the timber and branches are usually burned.... The second stage of deterioration is so-called orchard bush. Large trees still remain although widely scattered. The fringing forest slowly deteriorates into savannah. Sand dunes appear and where topsoil has been blown away the rock is exposed. The fierce heat of the sun bakes the rock until it explodes; as the temperature falls at night the rock, suddenly cooled, splits and cracks. These pieces of rock are ground together by the wind until they pulverize into even more sand.

The winds, unchecked by tree cover, grow in strength until they pile the sand into huge dunes, accumulated around a rocky spur or hill. The sand becomes dry, so dry that the tread of man or beast will make the dunes audibly hum and finally thunder as the sand slips down to a lower level.

All of these conditions and stages are to be seen in the Sahara: even the forest fringe is being driven back by the encroaching desert.

The answer to the desert is to give back to the land the tree cover which has been taken away. Where trees will grow, land will live again. [101]

The deserts of the world are on the march. Civilization, so-called, has been ruthless in the destruction of natural resources, so that the very existence of man on his planet is now being threatened. Can the tide of destruction be stemmed? This is a question we are unable to answer. Nature pays her debts, and when she is disregarded exacts terrible penalties. [102]

As we look at the world today, we see many parts that have been denuded of tree cover. During this past century we have bitten deeper into the natural resources of the earth than all former generations of mankind. We have upset the water cycle by removing the tree cover.

Modern techiques have speeded up the process of destruction. It took about fifteen hundred years for the Arabs to make the Sahara Desert. In the United States it has taken only about forty-four years to form the Dust Bowl which [spread] very rapidly. [103]

The Ancients... believed that the film of growth around the earth was its protecting skin and that... this skin — its vegetative covering of forests and growth spread out over its surface is certainly sensitive to the elements. Light, heat, fire and water promote, retard or stop the growth of plants and vegetation, upon which the life and well-being of the earth depends. [104]

In Nigeria, groundnuts or peanuts have been grown successfully for a hundred years or more. The nuts are planted between trees and they provide their own water in a mysterious and wonderful way. Each morning the plants open out and their leaves cover a relatively large area of ground. In the evening the leaves close up again, leaving a cool patch of earth around them. The hot, moist air rushes in to fill up the cool space. It condenses as dew; the cool, glistening droplets of water giving the equivalent of a quarter-inch of rain each night.

This wonderful plan of nature will only function when surrounding trees provide relatively cool conditions during the day in the upper air and give off enough moisture to produce a dew fall.

When the British government launched their groundnut scheme this important fact was forgotten — or not even considered! Into this scheme we rushed as madly and blunderingly as the machines which finally produced a new desert. The first onslaught was made by bulldozers. Sometimes half a dozen of these monsters were chained together and driven relentlessly across the ground, carrying everything before them. Bush, scrub, trees — trees that would have produced conditions favourable to the nut-bearing plants, were pushed off the earth and burned. Finally the tree cover was completely destroyed.

The scheme was a complete failure and in the space of a few years man produced a new desert. There now stands a vast area of useless, rolling sand, two hundred miles long and ninety miles deep which cost the British taxpayer £36,500,000. The expenditure of the money got quick results quite unexpected from those anticipated. Instead of plenty it brought catastrophe. The rhythm of the forest was broken, the delicate balance maintained by tree cover was destroyed and a new desert produced in shorter time than any other in the world. [105]

Nowhere would I permit clear-felling. I had seen the bad effects of this, not only in Equatorial Africa, but in Germany where the land had deteriorated as a result. I had wanted to save the English countryside from the risk of having the water table lowered and the land made feverish through opening it up to full exposure. Under proper management those who have to deal with the land will know the normal temperature, and it is important that it should be kept normal. When the earth covering is removed, the land becomes abnormal and feverish. It should always be the aim of foresters to prevent clear-felling and to provide timber by carrying out either stem selection or group selection felling. [106]

While on the Sahara expedition: I was anxious to get to Kenya. Thirty years had passed since I worked there as Assistant Conservator of Forests. Among the Kikuyu, I had first founded the society of the Men of the Trees.... When I reached them, I could try to find out what had happened to them in recent years. But Kenya was still many, many miles away.

After 2,400 miles of desert crossing, we entered Kano *[city in Nigeria]*, the back door to Equatorial Africa. We were fortunate to have completed the most arduous part of our journey without accident, no engine trouble, not even a puncture, though we had passed so many wrecks on the way.

We had taken a calculated risk and won through. We had charted remains of ancient forests, collected many botanical specimens, and estimated the speed at which our enemy, the desert, was invading the food-bearing lands of Equatorial Africa.

The day before entering Kano we had crossd the desert fringe and it had been terrifying to see the speed at which it was advancing. The wind-blown sand was burying crops before the peasants had time to harvest them. As far as we could see, there was no restriction on felling and by clearing the forest to make their farms, the people were clearing themselves. [107]

You can't blame the Brazilian government altogether for the decimation of their forests.... But the land they are using is quite unsuited to agriculture. In fact, it is quite unsuited for growing grass. They can probably grow grass for only one or two years at most; then the land erodes away because there are no tree roots to keep it intact. The soil starts to wash away as soon as the evergreen forest is removed.... But what's the rest of the world going to do for its oxygen? [108]

I look at it in this way. On our station at Mt. Cook, New Zealand, where I live, we never dream of using fertilizer, and my brother-in-law gets the top price for his Moreno wool. It's good management and you mustn't crowd that land with sheep. The Moreno likes to wander over a wide area. It takes ten acres to support one sheep, which seems a lot, but it's not sheep country really, it's forest country. My father-in-law planted half-a-million trees as shelter belts fifty years ago and those trees are now 135 feet high and four feet in diameter. I've worked out the wood increment and I find they pay fifty times better than wool, a rent of five dollars per acre per year compared with ten cents for producing wool. [109]

I've observed while attending World Forestry Congresses that more and more, my science, the science of silviculture, has been prostituted to short term economics. Silvilculture is so focused on the end product, paper pulp and timber, that it loses sight of the many vital values of a healthy forest. It's all short term and large scale clear-felling. [110]

The conquest of the desert will have to start with the conquest of the desert of the heart of man. We have witnessed tremendous strides in scientific research and invention, but it is obvious that the spiritual advance of mankind has not kept pace with scientific progress. [111]

The reclamation of the Sahara is one of the biggest tasks ever faced by men. Here is a desert more than one and a half times the size of Australia. We wish to restore its fruitfulness. Not in my time, but some day, there may be farms on those wastes; orchards, gardens and great cities and communities of men and women. Under the sands, nature has secreted its reserves; they must be freed for the benefit of mankind. We have the knowledge; we must find the means. [112]

It is not even a project for one nation. Many nations, all striving together with all the resources at their command, may be able to do something. Time and natural laws can be as much their aid as their enemy.

Perhaps the complete overall plan of the Sahara is a greater one and may take longer, both because of its vast size and because of its climatic factors: but if a belt a thousand miles long and one hundred miles wide can be planted in the U.S.A., and a three-thousand-mile shelter belt can be established in the U.S.S.R., and the Great Wall of China, two thousand miles long, can become a wall of trees to keep out the desert, why cannot a belt four thousand miles long and thirty miles wide be planted in Africa?

All through the present age of industrialism, man has been trying to conquer nature and now the implacable forces of the natural cycle are hitting back. The deserts are advancing. The Sahara is advancing. Man must accept its challenge. [113]

The basic plan in the Sahara Reclamation Programme is to divert military and other personnel and... funds from the more than 120 billion dollars spent yearly on armaments, to carry out large scale reforestation in the Sahara to modify the climate and make food production possible over ever-widening areas.... This most famous desert of the world is a vast reservoir of lands and natural resources.... Africa has a far greater population, already 230 million with a preponderance of youth, forty-five percent being under fourteen years of age. Africa is an important new element in the United Nations not only on account of the votes it represents but culturally because it has an entirely new and as yet unforeseeable contribution to make to world civilization.

When the newly liberated states unite to form the United States of the Sahara to help in providing food for a hundred million members of the human family now threatened by starvation, it will be the turn for the other countries to rub their eyes, stretch and bestir themselves, and play their part with the New Africa in saving humanity from the scourge of starvation. The goodwill of 230 million African peasants, herdsmen and fishermen will be a substantial element in the world of tomorrow and perhaps carry a greater weight than the 20 million or so town dwellers with their more sophisticated outlook somewhat restricted by political formulae, based on the already outworn attitudes of the West. [114]

The fact that we have a Sahara problem is not entirely tragic. The very existence of the Sahara gives to the whole world a highly valuable lesson in ecology. It teaches us what not to do with a perfect countryside. The drifting sands and stony wastes tell us more eloquently than words, what will happen when we break certain natural laws. We *cannot* remove tree cover without running the risk of losing the blessings of the water cycle. We *cannot* denude the earth's surface without creating the desiccation of sand and dust dunes. We *cannot* permit animals to devour whatever little is left of green growth. Excessive grazing of cattle, sheep and goats is as damaging to the land as is wholesale felling of trees....

The Sahara affords a challenge not to Africa alone but to the world. It offers a glorious opportunity to sink our differences and past exploitation of our planet into a new creative economy — a new way of living — a biologically correct way of life — bringing health and security in place of the anti-social scramble for money and power which can only end in a world holocaust. The task is a gigantic one and it cannot be achieved over night. The plan that we have outlined is both ethical and practical but will undoubtedly call for much individual sacrifice. However, the alternative is death — death to the earth beneath our feet, death to the neighbour by our side, and the end of all we love and hold dear. Surely no sacrifice will be too great if it will both save our planet from annihilation and bring a new conception of our oneness and interdependence under a Divine Providence. [115]

8

History Repeats

"The rise and fall of known and unknown civilizations waxed and waned as they exploited and devastated their forests."

History Repeats

For St. Barbe, the history of civilizations and the benchmarks within his own ninety-three years of experience provided ample evidence of the folly of man in relation to the earth and its cover. Into his writings, he weaves the history of forests in different cultures, to show how men have unknowingly and repeatedly devastated their own civilization by removing the precious green and humus-laden sponge from their surroundings.

A mighty demon of destruction has been loosed in the world during the past fifty years. The tremendous material strides that have been made by our modern civilization have eaten into the natural resources of our planet. The impact of modern industrialism, with its insatiable appetite for raw materials, caught the forests of the world before man had become aware of his eternal dependence upon them. To the trees he owed the stored-up wealth of the coal beds, the fertile earth for the production of food, water for irrigation, and the very air that he breathed. The rise and fall of known and unknown civilizations waxed and waned as they exploited and devastated their forests. Such is the story of the ancient civilizations of China and the countries surrounding the Mediterranean. The Golden Age of Africa was at its peak before the Sahara was a desert. When the Indian freely roamed the continent of America, there were some 820,000,000 acres of forest land. Much of this has been cut for farms and pasture, for cities and suburbs, as the population has grown. [116]

Speaking broadly, it has never happened in history that sylvan economy has been maintained. The story of... past civilizations has coincided with the exploitation of the forests — each rose to their peak, then declined as green forest mantle was lost. Unaware of their dependence upon trees, they failed to preserve a reasonable soil cover, and so exposed their society to the fall which must come as the result of such ignorant negligence. [117]

The historian of the future, looking back on the years from the invention... of the steam engine to the growing dominance of the jet-propelled engine, will have much to arrest his attention. He will be struck with the immense strides in engineering and the physical sciences and he will ponder upon our almost implicit belief that we are all sharing in something called *progress.*

Multitudes of people have worked and are still working away under the impression that the inevitable rewards are being brought nearer, when all the time they are rapidly undoing all that the unaided earth itself had built up in far-off ages. [118]

What... happened to the great eucalyptus forests of Australia? Those virgin domains were protected for countless generations by the inoffensive aborigines who... had been content with dried sticks to cook their food.

Into this forest paradise, with its two hundred and thirty-five species of eucalyptus, colonists came. The rhythm of growth was broken by wholesale felling, when wide stetches of forest were laid bare to grow crops.

To them the bush was there only to be attacked with axe and fire. They found, after a very short time, that the ash from the burned trees formed a hotbed for their seed, and more and more forest was laid low.

Not content with the felling of single trees, they invented an ingenious and infamous method by which they could clear an acre in less time than it took to fell one tree.

The axeman lays into the trees in such a manner that each falls upon its neighbour and takes it with it. This saves labour, especially in hilly country. When they are set off by felling of the big drive trees — those high up on the hill — the trees below the *drive* fall like ninepins until the mountainside looks like the aftermath of a tornado. Trees are shattered and scattered in all directions. Timber that would have made thousands of homes and provided shelter for generations is torn to splinters; while the mountains echo with the sounds of explosions like an artillery bombardment, as the stricken trunks fall to earth.

Tall timber which had clothed the mountain for centuries is laid low and left in a tangled mass. By the end of the day the ground is covered to a depth of twenty or thirty feet with the debris of splintered trees and branches.... It is left until it becomes as dry as tinder.

Then oil-soaked sacks are laid on the dense dry branches and lit. Flames roar up the mountain sides. Black clouds of smoke rise over the area and then go higher with the fierce heat that is generated. The inflammable oils from the eucalyptus produce highly combustible gases. Furnace heat is generated and soon the whole valley ignites with deafening explosions, blowing burning trees into the air and hurling them like fiery javelins across the valley to spread the conflagration. The... wealth of a million years in humus and timber has gone into the inferno. Rocks explode with the heat like land mines in some awful air raid. The burning forest travels faster and faster as this fiery, man-made hurricane is whipped up by the growing intensity of heat; flaming logs and tree trunks are driven before its force like leaves in an autumn gale.

The speed of destruction is that of an express train. When the fire dies down, the seed is quickly planted, even while the earth is hot and sometimes still smouldering. Many planters have been badly burned through working while the ground was still alight. This gigantic hotbed produces incredibly quick crops of wheat. The process starts all over again the next year. [119]

Whenever man has sought for land to grow food, he has found it in the forest. [120]

North America, with one-twelfth of the world's inhabitants, uses nearly half of all the timber and more than half of all the sawtimber. Its *per capita* consumption, one hundred and eighty-eight cubic feet, is five times as great as that of Europe. [*circa 1943.*]

Where are the forests going? Estimates show that America uses in one form or another about twenty-three billion cubic feet of wood every year. This means that about two hundred and fifty million trees of average size are cut from the forests every year ...trees that would cover eighteen thousand, five hundred square miles, or an area equal to all of England. Forest fires, decay, and insects destroy annually about two billion feet more. [121]

In years to come this profligate exploitation of land *[in the North American continent]* will no doubt be regarded as sheer madness, and yet to the people who were the actual cause of it — namely, the lumberers, farmers, and ranchers — it did not appear to be robbery and wastage of the national wealth.

The fault lies chiefly in our manner of translating wealth into financial figures, and then manipulating the financial figures to estimate the play of social forces. Lumberers and farmers had to pay their way, and the more money they made, the greater was the social esteem in which they were held. Paying your way and making money were perhaps the greatest recognized virtues at the period which brought about the disaster in natural wealth. To the early settlers, it certainly never occurred that the time would come when maltreatment of soil would ever be noticeable in such a vast territory. To them the new land appeared utterly inexhaustible. But their example followed by millions of successors, told a different story. [122]

Three hundred years ago settlers from England arrived on the

North American continent and established themselves at Plymouth and at Jamestown. Around them was nothing but forest. They cut trees to make log cabins and lived on the natural foods of the forest.... To these settlers the forest was an enemy to be overcome, and they waged ruthless war upon the trees. From the earliest days of their settlement, the home country called upon them for pine and spruce masts for her ships, and for fine oak timber for the hulls of Britain's growing navy. This was the beginning of the lumber industry. For a hundred years the white pine of the New England forests was sacrificed to build the houses of the growing colony. This was the beginning of the reckless career of lumbering which exposed the land to erosion and left ruined townships in its wake.

While the lumber magnates were turning the virgin forests into dollars, the farmers of the Middle West were growing wheat to satisfy the needs of the population which had sprung up on the wealth of the woods. The Industrial Revolution in England gave rise to a demand for cheap food and this led to further plowing of the prairies of the Middle West. Never before in the history of the world had wheat been produced with so little labor; crop after crop was extracted from the deep rich loam of those farmlands, until the soil became exhausted, and was then abandoned for fresh virgin lands to the west. Here was repetition of the nomadic farmers of Africa, only this time it was on a gigantic scale. Instead of the man with the machete, it was the man with a plowing outfit: first with his gang plow and six horses, breaking four and one-half acres of virgin prairie each day, then with a tractor, enormously increasing his speed in exploitation — but always leaving derelict land behind him. It was a rake's progress which spelled inevitable ruin.

Such exploitation of the soil deprives it of its protective covering. The land was denuded of trees, scorched by burning, exhausted by overgrazing of the herbage, deprived of its humus by constant cropping without replenishment, and inevitable disaster followed.... During the dry season the ground surface was reduced to dust and swept along by the wind. Without forest protection, these winds gathered force and played havoc with the denuded lands. When the rain came it was of little value, for the dust could not absorb it, and what remained of the soil with its valuable salts, was washed along the gullies to the flooding rivers.... At the time of World War I, wheat fetched peak prices, and every available acre of land, whether suitable or not, was plowed for grain. After the war, when prices dropped, whole farmlands were abandoned, and being uncultivated, were at the mercy of the elements. Standing one day in Chicago... choking with dust, [I asked] my companion the cause of this unpleasant experience [and] received the reply: "Oh, that's the Dakotas." This dust was what remained of the topsoil of once rich wheatlands and was blown by the wind for hundreds of miles, not only causing discomfort in the cities to the east, but baring the rocks from which it came. [123]

Referring to land destruction in Alberta: Coming from the last fertile country of the west, I met hundreds of distracted farmers from the United States. They had come from the newly formed Dust Bowl. Many of their farms had been buried by drifting sand, or the topsoil had blown away, making it impossible for them to grow any sort of crop. They had heard of the deep, rich black loam of Northwestern Alberta, and were moving north to continue their nefarious work of wheat growing and ultimate destruction. They told me that in many of the worst-hit states nearly all the farmers were getting out. After perhaps three successive crop failures, the farmers could not stand it any longer. I listened to their bitter humor, knowing they were unconscious that they themselves had wrought the destruction and were fleeing from a hell of their own making. [124]

Greece was another country which in times gone by was richly clothed with evergreens. Originally the whole of this peninsula, with a few small exceptions, was a continuous forest. Under the regime of the ancient Greeks the forests were protected by thousands of woodland deities and nymphs of the holy groves, until the fanaticism of the early Christians led to a war against these pagan strongholds in which [they] were destroyed by ax and fire. Centuries of misrule, overtaxation, reckless cutting, and extensive herding of goats and sheep, together with fires... reduced the forest area until today it covers but 1,917,980 hectares, or 15.1 per cent, of the whole area of that country *[circa 1940.]* Indeed, there is no virgin forest left. The many islands are entirely deforested and so are the seashores — where in olden times dense shady poplars stood — leaving a scene of barren sand and dreary rock. [125]

As populations grow and as living standards become higher, human wants become more complex and timber consumption increases. For every single substitute for wood in the world today about ten new uses are found for forest products in one form or another. Before World War I there were four hundred known uses for wood. At the end of that war there were four thousand known uses for forest products.

The amount of wood grown in the world has been roughly estimated at 38 billion cubic feet per year. Now, bearing in mind that the consumption is 56 billion cubic feet, that shows that we are cutting, or using, at any rate destroying, 18 billion cubic feet per year in excess of growth. *[circa 1956.]*

The total value of the world's export of all forest products is in the neighbourhood of six percent of the total foreign trade of the world. This is increasing, and increasing very rapidly, especially in the departments of wood pulp, pressed wood, kraft and other woodland products such as plywood — substitutes for timber. *[circa 1956.]* [126]

A native Kikuyu, speaks to St. Barbe, circa 1953: "You remember me, Muthungu. You well remember me because I was at Muguga, on the nearest land to your tree-farm. I am farming adviser to my people. What I am going to tell you I want you to put into your head. This is the reason why people look so dull. They are left to live in a very small place. They are surrounded by Europeans on all sides, and the Europeans have taken all the land that they could be cultivating now. They have taken away their land at Dagoretti and so they only have a small place to live in. Children are being born, the population is increasing: and they are like grass which when burned down never grows. Young men are not as strong as they used to be in the olden days. In the old days people used to work their land on rotation. There is no room for that now. They are prevented from using this system of cultivation. They look dull, because they are being chased away from squatting up country, and coming into a small space, for large areas of land are full of coffee planted by Europeans. They are closely packed together and so that is why they are thin."

He picked four long stalks of grass and laid them on the ground parallel to each other. Pointing to the space between the first two he said: "My great-grandfather, my grandfather, my father and myself, have farmed that area for that time. It is about fifteen acres, and during all that time the land has supported four families, about eighty people. But now it is used up and will grow little of any good and we have no other place to grow food." [127]

In every quarter of the world we find the ruins of once flourishing cities — the decayed remains of great civilizations. Only too often are these remnants of the past discovered in the middle of bare, pitiless deserts.

In North Africa, where the creeping sand is now man's deadliest enemy, there once spread the rich lands of Carthage. In Persia we find a history of soil exhaustion, crop failure and ultimate abandonment of the land. The mighty Tigris flows through Mesopotamia along its elevated bed of eroded soil. Its course is marked by bare and foreboding scenes; by humps of sandy, barren soil, by ruins and desolation — all that remain of ancient cities.

In Central America, Ceylon and Colombo, we find traces of civilizations that vanished long ago. In these climates nature is more complacent, but man cut the trees and broke the natural water cycle. In tropical countries such as these, danger comes from the rains. Without trees, there is nothing to hold the soil and the swirling waters rush down from the hills and sweep the soil away forever. [128]

17 A photograph of California redwoods taken by St. Barbe, and which appeared in *Dance of the Trees*, published in 1956.

18 St. Barbe with a planting team of Beijing school children in front of the Chinese Academy of Forestry in 1981.

19 Covers of two of the more than thirty books which St. Barbe published during his lifetime.

20 An example of the *Tree Lover's Calendar* which St. Barbe produced for several decades.

21 In the California redwoods in 1982, visiting the Grove of Understanding.

22 Visiting a school for handicapped children in New Zealand in 1981.

Notes from the Sahara Expedition: On arriving at the little hotel we had chosen for our meal, we were astounded to find that the story of our peach stones had gone ahead of us through the French papers and illustrated journals. Our host anxiously enquired if we would give him some to try in his irrigated garden. This we readily did, and we exchanged some of our potatoes for dates and the well-cooked meal that we enjoyed. Potatoes were a luxury so far from the coast.

Near where the peach trees were planted we were shown the remains of a huge fossilized tree stump which had been brought in from the desert nearby; it was an unidentified species although a better specimen than any we had seen in the museum, and unlike any we had previously seen in the Atlas Mountains. We were now hundreds of miles from these mountains and it would be interesting to know how long ago this tree was growing and the extent of this forest. [129]

All day long we travelled towards mountains of solid rock, and in the evening we rested at the well of Ekker which is under the shade of a great tamarisk tree.

The resthouse keeper, a man of about fifty, was an excellent host and spoke good French. When he gave us some of his precious sticks to cook our evening meal on, we asked him where they came from.

"I bought them," he told us. "They still find them here and there, growing along the banks of the *oueds*" *[dried-out riverbeds or valleys].*

We were interested. We told him we were foresters and were looking for old forests.

"Why, when I was a boy," he exclaimed, "the whole of that hill up there," he pointed to the mountainside, "was a great forest."

This was thrilling news indeed! Here at last was the sort of evidence we were looking for. We had found a man who could give us the firsthand evidence that we needed. He had seen these things himself.

Eagerly we questioned him further.

He told us that when he was a boy there were crops in plenty.

This great forest was then known as the *Last Forest*; for his father had remembered others. But people had come and cleared the trees, even removing the stumps and roots from the ground for charcoal and fuel.

Next morning I was up at dawn and walked by myself toward the mountain. At its foot was dark sand. Something different from stone suddenly caught my eye. I picked it up. It was a bleached chip of wood. I soon found others all over the place, scattered in the sand and usually partially buried. I returned to the resthouse keeper with my finds.

"You'll find plenty of those all over the desert. They are not big enough for people to bother with," he said in a matter-of-fact way.

For us they were treasures, rewarding us for all the roughness of the journey; for us they were proof that in the heart of the Sahara, within living memory, the last forest had been cut down.[130]

Describing an area in French West Africa, near Agades: Here was some cultivation, but the fields were already being buried by the invading wind-blown sands. This area had withstood the pressure of the greater Sahara up to a few years before when it was still forest. The peasant farmers found fresh virgin land each year by clearing the trees. Now having cut down their protection they were in the last wedge of forest land, completely surrounded by the greater desert and cut off from all habitation or other source of food except the precarious crops which they were harvesting before moving on....

For hundreds of miles we had been passing through a graveyard of dying races. That solitary fallen tree we had recorded indicated the extent and speed at which the Saharan octopus has spread itself. It was hard to believe that when Livingstone was exploring Nyasaland, this area was part of the great rain forests of Nigeria. In other areas, flourishing farms have disappeared within the memory of man, and later during this expedition I was to trudge through sand wastes which had been my forest haunts when I had been in Africa thirty years ago. Here one could actually see all the process of degradation, from high forest through the stages of orchard-bush and savannah to drifting sand. [131]

On returning to Kenya, thirty years later: In the days when I was assistant Conservator of Forests, Elgon was not included in my territories, and I had long wanted to go and see it. At that time the indigenous forest still remained intact and I wished it could have covered a wider area, because this was an important water catchment area.

Upon it depended the fertility of extensive farmlands, now occupied by settlers. Now, *[thirty years later]*, I looked over a landscape which from all time had been forest but which in so short a while had been replaced by coffee and wattle plantations and extensive wheat fields. I was soon to discover that this part of the country was an oasis by comparison with many areas which were suffering from steady desiccation.

There is a school of wishful thinking which loosely talks of wet and dry cycles and would like to think that the present period through which we are passing is just a long dry cycle. I was soon to see alarming evidence that even this still-fertile part of Africa is not escaping the general drying-out process; for example, Lake Naivasha has receded about half a mile from where it was when I knew it thirty years ago. I drove over plateaux which thirty years before were well watered but which today had dried up and become semi-arid.

Many old friends who had now settled, and some who had been there in the old days, welcomed me. I went from farm to farm. Always I was greeted with the same complaints. The country was drying up, the rains were becoming irregular; the last short rains had failed completely. If even they who lived there had observed the progressive drying up, it was hardly astonishing that I who had been away for thiry years was shocked by it. [132]

F.D.R. told me that when he got back to New York he sought out Gifford Pinchot [*creator of the first U.S. Forestry Service*], who pulled out of his pocket two pictures: one was of a painting made in China 300 years ago, and the other was a photograph taken of the same site that very year. In the first, the hills were completely covered with an evergreen forest. At the base of the hill there was a populous city, settled in a fertile valley with a river running through it. The only indication of logging to be seen in the painting was a little water chute with some logs coming down the slipway. The photograph taken three hundred years later showed the same contour of the hills, but they were bleak and bare. Not a vestige of their former glory remained. The one-time fertile valley was strewn with great boulders, the cover gone, and the dried-up soil with its fertility lost forever. All that remained of that flourishing city was a few derelict huts. Mr. Roosevelt said that had made such an impression on him that... "from that time, I have never looked back. You know what I have done." [133]

9

New Earth

"The present is full of opportunity."

New Earth

St. Barbe called for a new relationship to nature, and one shared by the indigenous peoples of the world: a respect for all living things and an appreciation of our interrelatedness to all of life.

As a man of action, he wanted to see respect translated into specific reclamation efforts. Despite the scale of the environmental problems, he felt that the solution was for every man, woman and child to plant and maintain trees.

It is not merely that the world is bettered by saving, replacing and multiplying trees, it is that an aim of this kind becomes an impulse towards developing a mood and an outlook which will increasingly [make it] natural to think for the future, for other people, for generations yet unborn. Planting a tree is a symbol of a looking-forward kind of action; looking forward, yet not too distantly.[134]

Your love of your country will extend to a feeling for the earth beneath your feet; like the ancients, you will come to regard the earth as a sentient being. You will feel your kinship with the earth and wish to protect its waters from abuse. You will not exploit them to their detriment by making them plunge through hydro-electric turbines, but wish to restore to them their natural courses with their bends, narrows, and broads. You will develop a sense of the oneness of all living things and realize your dependence upon them. Your growing respect for the earth will check the abuse of chemicals, both in the destruction of so-called weeds and in the unnatural forcing of your crops. You will learn the art of building up the humus and of leaving the earth a little better for your having lived on it.

You will become wise, not only by book-learning, but by being able to read the book of nature. [135]

Nature may be wooed but not coerced. And so it is with human beings — they too must be wooed from their destructive habits and be made aware of their inward kinship with the earth and all living things. [136]

I have been named an *earth healer*. It is a proud title, one that inspires humility and an awareness of all a man owes to the past and to his friends and teachers. The influences that shape and largely govern his development are many, not all of them recognized and few fully appreciated. As I have shown, the love of trees was part of my life from the beginning. My first teacher was my father. But maybe my feeling for the forest and its life was quickened most by my friends of the Northwest frontier, the Cree Indians who lived on the fat of the forests, yet understood the trees, survived their perils and felt a part of their life. [137]

It is interesting to note that in the story of the Maori *[indigenous peoples of New Zealand]* creation, trees came first. Before felling a tree, a Maori asks permission from the spirit of the tree, and afterwards covers up the stump with foliage to protect it from harm. In the hands of such people it is no wonder that the original forests were so well preserved. [138]

When I got to know them *[the Kikuyu, a tribe in Kenya]*... I studied their tree lore and folk lore and soon entered into the secret of their lives, which from birth to death were dedicated to the worship of N'gai, the father God. They lived intimately with nature, and, like the Maoris, felt their kinship with all living things.

When they cleared a forest to make farms, they left great trees which to them were sacred and which collected to themselves the spirits of the smaller trees that they had to fell in the course of making a sufficiently large area in which to grow their food.

It was a revelation to me that these so-called primitive folk had a deep feeling for the trees, which were to them as their elder brethren. [139]

The oak worship of the Druids is familiar to everyone, and their old word for sanctuary seems to be identical in origin and meaning with the Latin *Nemus*, a grove or woodland glade. Sacred groves were common among the ancient Germans and severe were the penalties exacted for damaging the bark of growing trees or felling protective groves. [140]

The ancients believed that the earth was a sentient being and felt the behavior of mankind upon it. As we have no proof to the contrary, it might be as well to accept this point of view and act accordingly. [141]

The fate of an individual or a nation will always be determined by the degree of his or its harmony with the forces and laws of nature and the universe.... The fullness of life depends upon man's harmony with the totality of the natural cosmic laws. Our individual evolution is a job that has to be carried on day by day by each individual. It is a lifelong task. [142]

Through the trees, we equip ourselves and serve them and, by serving them, bring new life to our planet. [143]

A planter of trees is rendering a great service to his country and neighbourhood, for while those trees are growing, they are paying a rich dividend in the form of moisture precipitated and conserved, climate modified and the soil protected and enriched. The people of the towns, too, benefit from the trees which directly or indirectly produce water, light and heat. In addition to their beauty and grace, they bless the landscape with their fair presences.

The commercial idea, that the only good tree is a dead one, that is, only fit for timber, no longer prevails. The tree has much more than a commercial value. Its influence on the climate and water supply has come to be regarded as a subject worthy of study and consideration. [144]

The science of forestry arose from the recognition of universal need. It embodies the spirit of service to mankind, in attempting to provide a means of supplying forever a necessity of life, and in addition, ministering to man's aesthetic tastes and recreational interests. Besides, the spiritual side of human nature needs the refreshing inspiration which comes from trees and woodlands. If a nation saves its trees, the trees will save the nation. And nations as well as tribes may be brought together in this great movement, based on the ideal of beautifying the world by the cultivation of one of God's loveliest creations — the tree. [145]

As man is dependent upon trees, so trees are dependent upon man. This is now understood as is evidenced by formation of UNESCO and the World Forestry Charter gatherings convened by the Men of the Trees during the past eleven years. The time has now come for us all to become aware of our tree heritage which must be handed on for others to enjoy. [146]

A new generation must arise who, inspired by love of the earth, will work and replenish where their fathers worked and exploited.[147]

Today it is the duty of every thinking being to live and to serve not only his own day and generation, but also generations unborn, by helping to restore and maintain the green glory of the forests of the earth. [148]

All people should demand that reasonable tree cover be maintained, so that erosion and rapid runoff may be prevented; otherwise their health and well-being are imperiled. Every man, woman, and child is really under duress not to waste water, or use it without thought for nature's own needs. As people learn more about the essential tree cover and appreciate the increasing value of forests for watershed protection, they will see to it that the forests are not maltreated. Forest fires, which are largely the result of carelessness or thoughtlessness, continue to do great damage, threatening to deplete and reduce the tree cover to a point at which it cannot adequately serve its function. By the prevention of forest fires, enormous annual losses to every interest and section of the community may be minimized. There should be adequate conservation laws in every country and governments should purchase suitable watershed, cutover, idle, and semi-arid lands for reforestation under scientific forestry supervision. All means possible should be employed to encourage the planting of trees and the creation of forests to provide sufficient cover, so that an approximation to the natural circulation of water may be restored.[149]

At the moment the real issue facing the world is not whether this or that political system survives, it is not merely a question of maintaining a national status quo. The real challenge is to humanity as a whole. Are we fit to live? Are we fit to exist on the earth? Can we unite to stem the oncoming tide of destruction which, by our folly, we have let loose on ourselves? The answer to these questions will decide the future of our race — the human race. The tasks which confront us are sufficiently great in themselves to need the thoughtful and concerted action of every country on this globe. Erosion must be checked; oncoming deserts must be stopped. Air and water pollution must be stopped. Land must be made fertile again with the help of trees of mixed species, and the earth once again be clothed in a green mantle of trees. The balance of nature must be restored. Paradise must be regained. [150]

I picture village communities of the future living in valleys protected by sheltering trees on the high ground. They will have fruit and nut orchards and live free from disease and enjoy leisure... living with a sense of their oneness with the earth and with all living things. [151]

When men set out to conquer nature they succeed, as the deserts they have created bear witness. They can also help nature to right herself. But first there must be the determination and availability of the necessary resources. [152]

It is one matter to admire a little tree and quite another to help preserve a great forest. In Kenya, many men spoke of their love of trees but very little was spent on forest conservation and re-planting. [153]

The real problem in... regions of abandoned land is not always immediate reforestation, but soil building, for that must precede the productive forest. The rebuilding of a forest soil involves soil surface phenomena. When there is insufficient litter cover, it is not porous enough to be capable of absorbing rain. [154]

When we are considering the reclamation of waste lands it would be as well to arrange that conservation and reclamation should proceed side by side. Any plan to restore desert places should be accompanied by steps to prevent degradation of existing vegetation. [155]

It was a noble conception of the President to plant a wall of millions of trees down the center of the United States, consisting of windbreaks extending one thousand miles, from the Canadian border down to the Panhandle of Texas. The wall consisted of a series of belts, one hundred miles wide comprised of drought-resisting species. This backbone of trees was intended to check the wind, which for years had carried away the soil of the farmlands, and also to serve as a trap for moisture. The trees planted were suited to the climate, and the shrubs on the outside rows acted as windbreaks. Even among the experts there were many who prophesied failure; but, in spite of adverse conditions, the entire appearance of the countryside is now being changed. The shelter belt strips have transformed the sandy desert and have broken the flat monotony of the plains. Helpful birds are once more returning and the farmers are again able to plant under the protection of the sheltering trees. [156]

Yet if the armies of the world, now numbering twenty-two million, could be redeployed in planting the desert, in eight years a hundred million people could be rehabilitated and supplied with protein-rich food grown from virgin sand.

The present is full of opportunity. Never before in the history of the planet has mankind been given the privileges and opportunites that are at his disposal today. A great light has been raised and is penetrating the darkness of the world, but alas, too many with dust-blinded eyes have yet to catch the vision. Some of us have. That is our privilege and our responsibility. [157]

Our future is in our children; the future of the world is in their hands and in the hands of their children. Let us train them wisely, and see that the understanding and appreciation of trees is part of their heritage. [158]

Boys and girls who are trained to look with understanding at trees will form tree fellowships, they will learn of the trees' struggle for existence; they will learn that trees are forever giving to man and of how dependent he is upon them for health, food, pure water, crops, rain, rivers and streams, birds, shade, pure air — in short, life and prosperity. [159]

The New Earth Charter, written by St. Barbe for the Men of the Trees:
We submit that without fair play to earth we cannot live physically; without fair play to neighbour, we cannot live socially; without fair play to better self, we cannot live individually.

We believe in the development of a fuller understanding of the true relationship between all forms of life in an endeavour to maintain a natural balance between minerals, vegetation, animals and mankind — man being primarily dependent on the vegetation of the earth for both food and clothing. In order to get food, clothes and shelter to enable us to live our bodily life on this earth, we must take care of the earth and especially, not meddle wantonly with the natural circulation of water, which meddling has been the cause of great loss of soil all over the globe. We must rightly return to earth the waste of whatever we take from the earth.

We submit that water must be a basic consideration in all our national and earth-wide forest programmes; streams and rivers must be restored to their natural motion, and floods and droughts must be eliminated. Forests and woodlands are intimately linked with biological, social and spiritual well-being. The minimum tree cover for safety is one third of the total land area. Every catchment area should have at least this proportion of tree cover made up of mixed species, including broad-leaf trees; monoculture in any form being injurious to the land.

We believe in the traditional ideal that our fields should be *fields of the woods,* by which is meant landscape farming of every valley and plain, with woodlands in high places, shelter belts, nut and fruit orchards (of mixed species) and hedgerow trees everywhere. [160]

In this book *[My Life, My Trees],* I have tried to record some of the numerous experiments of my life; indeed I am vividly aware that life is one continuous experiment. I would like to feel that what I have been permitted to achieve in conservation and as an *earth healer* through a lifelong planting of trees may encourage many others to dedicate their lives to the service of the earth. [161]

Sources

All of the quotes in this book are taken from the works of Richard St. Barbe Baker listed below, with the exception of selected interviews from Man of the Trees, *and as otherwise noted in the specific reference by number.*

Dance of the Trees. Oldbourne Press, London. 1956, 192 pp.

Green Glory. A.A. Wyn, Inc., New York. 1949, 253 pp.

Land of Tané. Lutterworth Press, London. 1956, 142 pp.

My Life, My Trees. Lutterworth Press, London. 1970, 180 pp. Reprinted by Findhorn Publications, The Park, Forres IV 36 OTZ, Scotland. 1970 and 1988, 167 pp.

New Earth Charter of the Men of the Trees. Richard St. Barbe Baker Collection, University of Saskatchewan Archives, Saskatoon, Saskatchewan, Canada.

Sahara Challenge. Lutterworth Press, London. 1954, 152 pp.

The Redwoods. Lindsay Drummond, Ltd., London. 1943, 95 pp.

Trees in the Environment. unpublished manuscript, Lutterworth Press, London. 1973, pp. 182.

Locke, Hugh, Ed. *Man of the Trees.* The Richard St. Barbe Baker Foundation, Canada. 1984, 32 pp.

Notes

1 *My Life, My Trees*, pp. 10-11

2 *Green Glory*, pp. 21-22

3 *Trees in the Environment*, p. 57

4 *Dance of the Trees*, p. 189-190

5 *Trees in the Environment*, pp. 3, 4

6 *Dance of the Trees*, pp. 190-191

7 *My Life, My Trees*, p. 113

8 *Land of Tané*, p. 36

9 *My Life, My Trees*, p. 43

10 *Land of Tané*, p. 56

11 *Sahara Challenge*, p. 114

12 *Green Glory*, p. 84

13 *Land of Tané*, p. 106

14 *Sahara Challenge*, pp. 120-121

15 *Sahara Challenge*, pp. 112-113

16 *Dance of the Trees*, p. 165

17 *Green Glory*, p. 190

18 *Land of Tané*, p. 123

19 *Green Glory*, pp. 192-193

20 *Sahara Challenge*, p. 126

21 *Green Glory*, pp. 104-105

22 *Land of Tané*, p. 108

23 *Man of the Trees*, pp. 7-8

24 *Man of the Trees*, p. 11

25 *Green Glory*, pp. 114-115

26 *Green Glory*, p. 237

27 *Green Glory*, p. 238

28 *Dance of the Trees*, pp. 185-186

29 *Man of the Trees*, p. 10

30 *Green Glory*, p. 222

31 *Man of the Trees*, pp. 4-5

32 *Green Glory*, p. 222

33 *Land of Tané*, p. 92

34 *Land of Tané*, p. 78

35 *Dance of the Trees*, p. 55

36 *Dance of the Trees*, pp. 181-183

37 *Land of Tané*, p. 95

38 *Green Glory*, p. 201

39 *Man of the Trees*, p. 10

40 *Man of the Trees*, pp. 12-13

41 *My Life, My Trees*, p. 167

42 *Green Glory*, p. 148

43 *The Redwoods*, p. 84

44 *Trees in the Environment*, p. 1

45 *Man of the Trees*, p. 13

46 *Land of Tané*, p. 36

47 *My Life, My Trees*, p. 55

48 *Land of Tané*, p. 97

49 *Land of Tané*, p. 96

50 *Dance of the Trees*, pp. 63-68

51 *Green Glory*, pp. 24-25

52 *Land of Tané*, pp. 96-97

53 *Green Glory*, pp. 52-55

54 *Dance of the Trees*, pp. 72-73

55 *Trees in the Environment*, p. 6

56 *Dance of the Trees*, pp. 71-72

57 *My Life, My Trees*, pp. 166-167

58 *My Life, My Trees*, p. 56

59 *Green Glory*, p. 5

60 *Green Glory*, pp. 82-83

61 *Man of the Trees*, p. 5

62 *Sahara Challenge*, pp. 72-73

63 *Green Glory*, p. 196

64 *Green Glory*, p. 25

65 *My Life, My Trees*, pp. 55-56

66 *Land of Tane*, pp. 37-38

67 *My Life, My Trees*, p. 165

68 *Green Glory*, p. 78

69 *Green Glory*, pp. 195-196

70 *Green Glory*, p. 125

71 *Dance of the Trees*, p. 52

72 *Green Glory*, pp. 79-80

73 *My Life, My Trees*, p. 21

74 *The Redwoods*, p. 80

75 *Sahara Challenge*, p. 114

76 *Green Glory*, pp. 199-200

77 *Green Glory*, pp. 146-147

78 *Sahara Challenge*, pp. 35-36

79 *Sahara Challenge*, p. 83

80 *Sahara Challenge*, p. 93

81 *Land of Tané*, pp. 65-66, 68-69

82 *Dance of the Trees*, p. 121

83 *The Redwoods*, pp. 9-10

84 *The Redwoods*, p. 42

85 *Dance of the Trees*, pp. 113-114

86 *The Redwoods*, p. 36

87 *The Redwoods*, p. 88

88 *The Redwoods*, p. 5

89 *Land of Tané*, pp. 52-53

90 *The Redwoods*, pp. 58-61

91 *Dance of the Trees*, p.94

92 *My Life, My Trees*, p. 145

93 *The Redwoods*, p. 91

94 *Sahara Challenge*, pp. 96-97

95 *My Life, My Trees*, pp. 43-44

96 *Green Glory*, pp. 184-185

97 *Green Glory*, p. 26

98 *Sahara Challenge*, p. 24

99 *Land of Tané*, pp. 137-138

100 *My Life, My Trees*, pp. 162-163

101 *Dance of the Trees*, pp. 140-141

102 *Trees in the Environment*, p. 37

103 *Dance of the Trees*, pp. 68-69

104 *Sahara Challenge*, p. 19

105 *Dance of the Trees*, pp. 52-53

106 *Dance of the Trees*, p. 129

107 *Dance of the Trees*, pp. 152-153

108 *Man of the Trees*, p. 10

109 *Man of the Trees*, p. 11

110 *Man of the Trees*, p. 12

111 *Land of Tané*, p. 44

112 *Dance of the Trees*, p. 184

113 *Trees in the Environment*, p. 93

114 *Trees in the Environment*, p. 163

115 *Trees in the Environment*, p.168

116 *The Redwoods*, p. 79

117 *Green Glory*, p. 222

118 *Land of Tané*, pp. 44-45

119 *Dance of the Trees*, pp. 108-110

120 *Green Glory*, p. 23

121 *The Redwoods*, p. 83

122 *Green Glory*, p. 229

123 *Green Glory*, pp. 59, 62-63

124 *Green Glory*, p. 66

125 *Green Glory*, p. 129

126 *Dance of the Trees*, p. 134

127 *Sahara Challenge*, pp. 128-129

128 *Dance of the Trees*, pp. 51-52

129 *Sahara Challenge*, pp. 46-47

130 *Sahara Challenge*, pp. 53-54

131 *Sahara Challenge*, pp. 69-70

132 *Sahara Challenge*, pp. 110-111

133 *Green Glory*, p. 67

134 *Trees in the Environment*, p. 1

135 *Land of Tané*, p. 140-141

136 *Land of Tané*, p. 139

137 *Dance of the Trees*, p. 50

138 *Dance of the Trees*, p. 107

139 *Land of Tané*, pp. 86-87

140 *Green Glory*, p. 103

141 *Dance of the Trees*, p. 68

142 *My Life, My Trees*, p. 163

143 *Land of Tané*, p. 96

144 *Land of Tané*, p. 74

145 *Trees in the Environment*, p. 3

146 *Dance of the Trees*, p. 191

147 *Green Glory*, p. 196

148 *Green Glory*, p. 243

149 *Green Glory*, p. 238

150 *Trees in the Environment*, p. 30

151 *My Life, My Trees*, p. 167

152 *Dance of the Trees*, p. 59

153 *Dance of the Trees*, p. 34

154 *Green Glory*, p. 153

155 *Dance of the Trees*, p. 73

156 *Green Glory*, pp. 72-73

157 *My Life, My Trees*, p. 163

158 *Dance of the Trees*, p. 192

159 *Dance of the Trees*, p. 192

160 *The New Earth Charter of Men of the Trees*

161 *My Life, My Trees*, p. xv, Preface

Photographic Credits

1 Howard Coster

2 Reprinted from *Dance of the Trees*

3 Richard St. Barbe Baker Collection, University of Saskatchewan Archives

4 Richard St. Barbe Baker Collection, University of Saskatchewan Archives

5 Richard St. Barbe Baker Collection, University of Saskatchewan Archives

6 Richard St. Barbe Baker Collection, University of Saskatchewan Archives

7 Richard St. Barbe Baker Collection, University of Saskatchewan Archives

8 Howard Coster

9 Richard St. Barbe Baker Collection, University of Saskatchewan Archives

10 Richard St. Barbe Baker Collection, University of Saskatchewan Archives

11 Ted Cardell

12 Reprinted from *Sahara Challenge*

13 Reprinted from *Sahara Challenge*

14 Esso Photograph FA0611

15 Richard St. Barbe Baker Collection, University of Saskatchewan Archives

16 Richard St. Barbe Baker Collection, University of Saskatchewan Archives

17 Richard St. Barbe Baker

18 Chinese Academy of Forestry

19 Reprinted from the covers of *Green Glory* and *Sahara Challenge*

20 Richard St. Barbe Baker Collection, University of Saskatchewan Archives

21 Randy Stemler

22 Auckland Star